12 A

2

A geography of energy in the United Kingdom

A geography of energy in the United Kingdom

John Fernie
with a Foreword by Tony Benn

Longman
London and New York

To Suzanne

Longman Group Limited London

*Associated companies, branches and representatives
throughout the world*

*Published in the United States of America
by Longman Inc., New York*

© Longman Group Limited 1980

First published 1980

British Library Cataloguing in Publication Data

Fernie, John
  A geography of energy in the United Kingdom.
  1. Power resources – Great Britain
  2. Great Britain – Economic conditions – 1945–
  333.7   HD9502.G72   79–40858
  ISBN 0-582-30007-X

0 582 30007 X

Printed and Bound in Great Britain by
McCorquodale (Newton) Ltd, Newton-le-Willows, Lancashire

Contents

List of plates

List of figures

List of tables

Foreword

During recent years the Department of Energy has worked consistently to promote open public discussion on energy policy, to make widely available all relevant information and to draw into the process of energy policy formulation all those who have something to contribute to it. Energy policy is a very important subject. The decisions we take now will affect not only ourselves and our living standards but also those of our children, and the ever-growing public interest in energy matters reflects this fact. It is, therefore, important that these decisions should be made within the framework of a national energy strategy which is based on the widest public understanding and acceptance.

If the public debate on energy policy is to be truly representative of all the interests involved, it is necessary for all viewpoints to be expressed and considered. Much progress has already been made through the holding of the National Energy Conference in June 1976, and the establishment of the Energy Commission, whose papers are available to the public and to whom anyone may comment. While these are major steps forward, there is also a need for contributions to the debate from sources outside the Government. I therefore have great pleasure in welcoming the publication of Dr Fernie's book as introducing another viewpoint to the discussions. I am sure this will prove a valuable addition to the information available on energy in the UK.

Tony Benn

Preface

In the 1970s public interest has been focused on energy. The 'crisis' of 1973/74, the growing strength of OPEC and short-term shortages of oil because of the revolution in Iran have dramatically brought the period of cheap oil to an end and highlighted the political uncertainty of over-dependence on imported fuel.

In a quest for self-sufficiency, North Sea oil and gas is being developed, the expansion of coal production capacity is under way, whilst the nuclear industry could guarantee electricity supplies in the long term. What should be the 'energy mix' of fuels to supply anticipated demand? How many new coal mines, power stations and petrochemical complexes will be necessary in the future and where will they be located? Britain has an abundance of energy resources but their future development will radically affect areas as yet untouched by industrialisation. North Sea oil development has attracted manufacturing and service activity to unspoiled, beautiful areas of the Highlands and Islands of Scotland; the Selby coalfield will bring change to the rural landscape of North Yorkshire. The Belvoir public inquiry is only the first of the many major inquiries where the forecasts propounded by the fuel industries and endorsed by the Government will come under public scrutiny.

The aim of this book is to provide the geographer with an independent assessment of energy options open to the Government. From the evidence produced, the reader can make a judgement on desirable policies for the future. Following a review of energy policy since 1945, various scenarios for a long-term strategy are outlined. Each fuel and power industry is discussed in detail to assess its past, present and projected contribution to energy supplies. The potential of renewable energy resources and more efficient utilisation of energy through conservation measures are examined in the final chapters. Throughout, close attention is given to the environmental impact of existing and proposed projects.

The energy situation is dynamic. *A Geography of Energy in the UK* was completed by Christmas 1978. Since then, events such as the oil spill at Sullom Voe and the NCB proposals to re-open the Thorne colliery endorse or expand upon the author's arguments in the relevant sections. The Harrisburg incident in the USA may have, in addition, significant implications for UK energy policy. The prospect of a 'melt-down' or an explosion of a PWR at the Three Mile Island site will inevitably cause much controversy both inside and outside the nuclear industry.

Policy formulation will be increasingly influenced by public opinion. This book provides both the geographer and the layman with a wider understanding of the complex and controversial issues involved in energy planning in the UK.

I would like to express my thanks to the Department of Energy, the nationalised fuel industries and Fife Regional Council for all their help during the writing of the book. Special thanks to my wife Suzanne who read, amended and typed the manuscript and who became a book widow for much of 1978.

John Fernie
Huddersfield
March 1979

List of abbreviations

AGR	Advanced Gas Cooled Reactor
BNFL	British Nuclear Fuels Limited
BNOC	British National Oil Corporation
BP	British Petroleum Company
BWR	Boiling Water Reactor
CEGB	Central Electricity Generating Board
CHP	Combined Heat and Power
CIA	Central Intelligence Agency
COP	Coefficient of Performance
ERDA	Energy Research and Development Administration
ETSU	Energy Technology Support Unit
FBR	Fast Breeder Reactor
FOE	Friends of the Earth
GEC	General Electric Company
GGP	Gas Gathering Pipelines Ltd.
GNP	Gross National Product
GW	Gigawatts (1000 MW = Megawatts)
HEP	Hydro-electric power
HRS	Hydraulics Research Station
IAEA	International Atomic Energy Agency
ICRP	International Commission on Radiological Protection
ISES	International Solar Energy Society
JET	Joint European Torus
LNG	Liquid Natural Gas
LPG	Liquid Petroleum Gas
MW	Megawatts
NALGO	National Association of Local Government Officers
NASA	National Aeronautics and Space Administration
NCAT	National Centre for Alternative Technology
NCB	National Coal Board
NEDECO	Netherlands Engineering Consultants Foundation
NEDO	National Economic Development Office
NII	Nuclear Installations Inspectorate
NPT	Non-proliferation Treaty
NRPB	National Radiological Protection Board
NUM	National Union of Mineworkers
OECD	Organisation for Economic Cooperation and Development
OMISCO	Offshore Maintenance and Inspection Company
OPEC	Organisation of Petroleum Exporting Countries

OSO	Offshore Supplies Office
OTEC	Ocean Thermal Energy Conversion
PRT	Petroleum Revenue Tax
PSA	Property Services Agency
PWR	Pressurised Water Reactor
R and D	Research and Development
RDL	Redpath Dorman Long
RIBA	Royal Institute of British Architects
SGHWR	Steam Generating Heavy Water Reactor
SNG	Substitute Natural Gas
SNP	Scottish National Party
SSEB	South of Scotland Electricity Board
TCPA	Town and Country Planning Association
THORP	Thermal Oxide Reprocessing Plant
UKAEA	United Kingdom Atomic Energy Authority

Metrication table

1 kilometre	=	0.621 mile
1 metre	=	3.281 feet
1 centimetre	=	0.394 inch
1 square kilometre	=	0.386 square mile
		247.1 acres
1 hectare	=	2.471 acres
		1.196 square yards
1 square metre	=	10.764 square feet
1 metric tonne	=	0.984 short ton
1 kilogramme	=	2.205 pounds
1 cubic metre	=	35.315 cubic feet
1 cubic centimetre	=	0.061 cubic inch
1 litre	=	0.220 Imperial gallon

Acknowledgements

We are grateful to the following for permission to reproduce copyright material: Blackie and Son Ltd for our Fig 2.3 from *Energy and the Environment* by Lenihan and Fletcher; Elsevier's Applied Science Publishers Ltd for our Table 6.1 by H. J. Alkema and E. V. Newland from *Energy: From Surplus to Scarcity* edited by K. Inglis; ERA Technology Ltd for our Fig 5.1 from 'The Potentialities of Wind Power for Electricity Generation' by Golding and Stodhart © ERA Technology Ltd 1979; The Fife Regional Council for our Fig 2.5 'Lochore Meadows – Proposals for a Country Park' Report by Department of Physical Planning, the Fife Regional Council; Financial Times for our Table 3.4 by R. Dafter in *Financial Times*, 9-11-77; Financial Times and IPC Magazines Ltd for our Fig 3.4 appeared in *Financial Times*, 9-12-77 and p 264 *New Scientist* 21-8-79; Geological Society of London and the author, Sir P. Kent for our Fig 3.1 in *Journal of Geological Society* Vol 131/5, September 1975; Her Majesty's Stationery Office for our Table 3.8 from *Scottish Economic Bulletin* No 11, 1977; our Table 3.9 from *Scottish Economic Bulletin* No 11, 1978; our Table 3.10 from *Scottish Economic Bulletin* No 15, 1978; our Fig 4.1 from *British Industry Today: Energy* by Central Office of Information, 1977; our Fig 4.3 from *Energy Commission Paper* No 1 by Department of Energy; our Fig 5.2 from *Energy Commission Paper* No 21 by Department of Energy; our Fig 6.1 from *Energy Commission Paper* No 11 by Department of Energy; our Fig 6.2 and Table 2.6 from *Digest of UK Energy Statistics* by Department of Energy, 1977; our Fig 2.4 from *Offices and Services Industries* by Department of Industry, 1976; our Table 6.3 from *The New Prospectors* by Warren Spring Laboratory, 1976; our Table 2.8 from *The Stevens Report Planning Control over Mineral Working* by Sir R. Stevens, 1976; our Tables 3.1, 3.2, 3.3, 3.5 and 3.6 from *Development of the Oil and Gas Resources of the UK* by Department of Energy, 1978; our Table 3.7 from *Annual Reports* by Offshore Supplies Office; our Tables 5.1, 5.3 and an extract from *Energy Commission Paper* No 16 by Department of Energy, 1976; our Table 1.1 compiled from data in *Digest of United Kingdom Energy Statistics* by Department of Energy, 1977 and *Energy Trends* by Department of Energy, 1978; our Table 1.2 compiled from data in *Energy Policy Review* Paper No 22 by Department of Energy, 1977 and *Working Document on Energy Policy* Paper No 1 by Energy Commission, 1977; our Table 1.3 compiled from data in *Energy Policy Review* Paper No 22 by Department of Energy, 1977, *Working Document on Energy Policy* Paper No 1 by Energy Commission, 1977 and *Energy Policy: A Consultative Document* CMND 7101, 1978; our Tables 6.4 and 6.5 compiled from data in *Transport Statistics Great Britain 1965–75* and *Annual Abstract of Statistics*, reproduced with the permission of the Controller of Her Majesty's Stationery

Office; Her Majesty's Stationery Office and The National Coal Board for our Tables 2.3 and 2.5 compiled from data in *Digest of UK Energy Statistics* by Department of Energy, 1977 and *National Coal Board Reports and Accounts* 1966/77 and 1967/68; Institute of Energy and The National Coal Board for our Table 2.7 compiled from data in *Energy World* No 34, 1977 and National Coal Board Public Relations Office Information; Charles Knight & Co Ltd and Her Majesty's Stationery Office for our Table 6.2 compiled from data in *Waste Disposal Management and Practice* by J. Skitt, 1979 and *Waste Management Advisory Council* 1st Report, 1976; The National Coal Board for our Figs 2.1, 2.2 and 2.7; our Table 2.2 from *NCB Annual Reports and Accounts*, 1966/67; our Table 2.4 from *NCB Annual Reports and Accounts*, 1957 and 1966/67; our Table 2.1 from Plan for Coal, 1950; and two extracts from advertising features appearing in *Financial Times*, 4-2-76, 15-9-76 and *The Guardian* 29-10-76; Kogan Page Ltd and the editor, Ethel de Keyser for our Fig 3.6 from *European Offshore Oil and Gas Yearbook*, 1975/6; Penguin Books Ltd for Fig 10 p 88 (our Fig 4.2) 'The Nuclear Fuel Cycle' in *Nuclear Power* by Walter C. Patterson (Pelican Books, 1976) © Walter C. Patterson, 1976, reprinted by permission of Penguin Books Ltd; Reading University for our Fig 5.3 from an unpublished paper by P. Musgrove, © P. J. Musgrove; Royal Scottish Geographical Society for our Figs 3.2, 3.3 and 3.5 in *Scottish Geographical Magazine* Vol 93, 1977.

Whilst every effort has been made, we are unable to trace the copyright owner of our Fig 2.6 and Table 5.2 and would appreciate any information which would enable us to do so.

Chapter 1

The changing energy situation

The Yom Kippur war in 1973 was a significant political event in energy history. Within two days of the outbreak of war, Arab oil producers jointly agreed to reduce production levels and impose an embargo on exports to specific countries, including the United States. The use of oil as a political weapon brought a period of cheap oil to an end. Previous attempts at nationalisation or production cuts by producer governments[1] had only led to short-term crises which the oil majors averted because of the availability of supplies from other oil provinces. During the six-day Arab-Israeli war in 1967 and the closure of the Suez Canal, the oil companies could have increased production from the US alone by 4 million barrels a day if the need had arisen.[2] By 1973, this excess capacity had been eliminated, the US had become more dependent on Middle East imports[3] and Arab action resulted in a quadrupling of oil prices as consumers looked to other OPEC members who were willing to maintain or increase production levels.

The 'energy crisis' has aptly illustrated the need for a long-term strategy to protect countries such as Britain from the vagaries of political action and give them greater flexibility in their choice of energy supplies. Britain, which is well-endowed with energy resources, imported oil for 46 per cent of her energy requirements in 1973, when energy planning was based on competition of fuels in a free world market context. The policies of successive governments have assured the British public of fuels at competitive prices and the mix of supply industries has been geared to achieving this aim. Investment decisions affecting the nationalised fuel industries have been taken to guarantee supplies in the long term. The fear of a discrepancy between supply and demand and the creation of an 'energy gap' has shaped policy in the post-war period.

Towards an oil-based economy

A shortage of fuel in the 1950s led to the development of nuclear power in response to concern about an impending 'energy gap' which did not materialise. Instead, oil supplies which had been curtailed during the Suez crisis became more freely available during the late 1950s and 1960s. With the closure of the Suez canal in 1956, tankers were forced to make a detour of 9,600 miles around the African coast. To cut these additional transportation costs, and benefit from scale economies, a new generation of ships was constructed – the super-tankers.[4] In addition to technological improvements in transporting oil, new oil provinces were developed in North and West Africa because of their geographical proximity to West European markets.

The abundance of oil on the world market depressed prices. Indeed, the creation of OPEC in 1960 was a reaction against the reduction of 'posted' prices in 1959. The lack of solidarity of the cartel, which did not negotiate as a single body in the 1960s, allowed the oil companies to continue to produce, distribute and market their oil at competitive prices. The 'real' price of oil products fell by 50 per cent in the 1960s.[5] The situation was to change drastically in the first half of the 1970s as the more militant members of the cartel demanded higher prices and participation in existing oil concessions. In January 1973, OPEC and the oil companies agreed to phase in participation by producer governments from 25 per cent in 1973 to 51 per cent in 1982. This would give the companies nearly a decade to operate and market existing concessions whilst they gained experience of new 'marginal' oil provinces, such as the North Sea. The bargaining position of the oil companies was further weakened after the Arab action of 1973/74, and OPEC agreed to determine the future price of oil unilaterally. As a result of this action, the oil companies were obliged to develop North Sea reserves more rapidly than anticipated. Odell has argued that in these circumstances West European governments could have dictated their own terms to the oil industry.[6]

The role of the Government – pre-1973

This background provides the framework for energy planning in Britain. Throughout the post-war period, the Government has played a passive role and allowed free market forces to operate while adopting 'ad hoc' measures when they were thought necessary. For example, throughout the 1960s coal mining, suffering from the cost competitiveness of oil, received Government support to ensure an ordered run-down of the industry to minimise social and economic disruption in the affected areas.

The administrative and statutory guidelines of the nationalised fuel industries reinforce the concept of free competition between the main supply bodies; their aims being to supply coal, gas or electricity to consumer requirements at prices which cover production costs. As early as 1952 the Ridley Committee advocated a policy of coordination through competition between the fuel industries and proposed that a Joint Planning Board should be created to ensure cooperation in production.[7] The Committee's suggestions were never fully implemented and most Government decisions have tended to affect individual fuel industries when intervention was felt beneficial to the national interest. In the 1950s, when Britain was still dependent on coal for over 80 per cent of her energy requirements, policy was directed towards the NCB, initially to encourage an expansion of production, then to conserve coal and ultimately in the 1960s to rationalise the industry. The development of a nuclear power programme in the 1950s to guarantee the long-term availability of power supplies is one aspect of Government policy which did not undergo the 'cooperation through competition' criterion. The competitiveness of oil and the development of natural gas in the 1960s undermined the urgency to expand nuclear power production and by 1969 the proportion of oil consumed increased to 43 per cent of the total energy consumption, whilst coal consumption fell to around 50 per cent of the total (Table 1.1).

Table 1.1 UK inland energy consumption of primary fuels (excludes non-energy uses and energy for bunkers) (*Sources:* E. S. Simpson: *Coal and the Power Industries in Post-War Britain*, Longman 1966; Department of Energy, *Digest of United Kingdom Energy Statistics*, HMSO 1977; Department of Energy, *Energy Trends*, Feb. 1978)

	Coal (%)	Petroleum (%)	Natural Gas (%)	HEP & Nuclear (%)	Total (million tons of coal equivalent)
1950	94.5%	4.9%	—	0.6%	212.6
1960	82.6%	16.6%	—	0.8%	238.0
1969	50.4%	42.9%	2.9%	3.8%	320.3
1973	37.6%	46.5%	12.5%	3.4%	347.9
1975	37.0%	42.0%	17.0%	4.0%	319.7
1977	36.4%	40.3%	18.5%	4.8%	332.4

Although the availability of cheap Middle East oil greatly influenced energy policy in the 1960s, the National Plan of 1964 and two White Papers in 1965 and 1967 gave a preview of official strategies for future energy planning. The 1967 White Paper re-emphasised that a cheap fuel policy 'having regard to the whole range of relevant considerations – economic and social' – would continue to shape Britain's energy future.[8] Despite the creation of OPEC, it was not thought necessary to discriminate against the oil industry on grounds of security. Six years later this policy was shown to be wrong and the forecasts for 1975 were largely inapplicable. The 'energy crisis' has demonstrated that energy is no longer cheap or infinite. The UK was fortunate in 1973 in that supplies of North Sea gas were building up to act as a stop-gap until offshore oil production could assure the country of self-sufficiency in energy by the end of the decade. The length of time Britain will be a net exporter of energy is in dispute;[9] nevertheless, the country has an excellent opportunity to formulate a long-term strategy in the knowledge that supplies are assured in the short term.

The role of Government – post-1973

The Government has now adopted a more flexible approach to energy planning. Mr Benn's policy of 'open government' enabled all parties which can contribute to energy policy formulation to express their views through publications and conferences, the most important to date being the National Energy Conference in June 1976.[10] Prior to the conference, the Department of Energy published a discussion document showing seven scenarios for a possible long-term strategy.[11] In the same year, the Department conducted an energy policy review which was subsequently published in 1977 in its own energy paper series.[12] After the National Energy Conference and the obvious need for co-ordination in energy planning, Mr Benn created an Energy Commission to advise and assist him in the development of a national energy strategy. The Commission first met in November 1977 and has produced a series of papers related to short- and long-term energy forecasts. Its first paper – a working document on energy policy – formed the basis of the Green Paper published by the Government in 1978 to act as a consultative document for a future strategy.

The difficulty in accurate forcasting is shown from official estimates of supply and demand to the end of the century. Within one year, the demand forecasts of the Department of Energy and the Energy Commission have varied by 10 to 35 million tons in 1985 and 50 to 90 million tons by the year 2000 (Table 1.2). In terms of supply, forecasts for the year 2000 indicate that the Department of Energy is more willing to keep options open in relation to costs of competing fuels than the Energy Commission and the Government, which show a greater commitment to the coal and nuclear power options to meet energy requirements by the end of the century (Table 1.3).

Table 1.2 Forecasts of energy demand in 1985 and 2000 (million tons of coal equivalent) (*Sources:* Department of Energy, *Energy Policy Review, Energy Paper No. 22,* HMSO 1977; Energy Commission, Working Document on Energy Policy, *Energy Commission Paper No. 1,* Department of Energy 1977)

	1975	1985		2000	
		Dept of Energy	Energy Commission	Dept of Energy	Energy Commission
Demand (high growth)					
Energy uses	314	400	375	580	490
Non-energy and Bunkers	27	50	40	70	70
Total primary fuel demand	341	450	415	650	560
Demand (low growth)					
Energy uses	314	340	350	440	390
Non-energy and Bunkers	27	40	40	60	60
Total primary fuel demand	341	380	390	500	450

Table 1.3 Forecasts of energy supply in 2000 (million tons of coal equivalent) (*Sources:* *Department of Energy, *Energy Policy Review, Energy Paper No. 22,* HMSO 1977; †Energy Commission, Working Document on Energy Policy, *Energy Commission Paper No. 1,* Department of Energy 1977; ‡*Energy Policy: A Consultative Document,* Cmnd 7101, HMSO 1978)

Indigenous production	Department of Energy*	Energy Commission†	The Government‡
Coal	100–165	170	170
Natural Gas	40– 75	50– 90	50
Nuclear/Hydro	27–102	95	95
Oil	50–150	150	150
Renewable		10	
Total	215–490	475–515	465

The official strategy has been criticised because of the diversity of forecasts of future demand and the mix of fuels intended to meet the projected 'energy gap'. The National Council for Alternative Technology (NCAT) claims that energy consumption could fall by 2025 if the elimination of waste in the conversion, distribution and use of energy gained top priority in a future strategy.[13] The reduced demand could be met by a combination of coal and renewable resources as opposed to the official proposal to develop coal and nuclear power. A hybrid strategy proposed by the Royal Commission on Environmental Pollution does not discount the nuclear option but envisages a much greater contribution from renewable resources, waste materials and waste heat which in total could supply 50 to 60 million tons of coal equivalent by 2000.[14] Renewable resources are not included in the supply formulations in the Department of Energy's Policy Review or the Government's Green Paper. Both papers omit renewable resources in their calculations because of uncertainties over their potential contribution, although they acknowledge that 10 to 40 million tons of coal equivalent could be supplied.

Forman is concerned about the inflexibility of the Green Paper proposals and, in addition to arguing for greater efficiency in the conversion, transmission and use of energy, feels that planning should be framed to coordinate supply with demand. He comments on the Government's plans for electricity production as 'putting the nuclear cart before the electricity horse'.[15] The implication from the Green Paper is that the electricity market will have to grow to meet projections of nuclear generating capacity by 2000. He adds that if natural gas had not been discovered in significant quantities, Substitute Natural Gas (SNG) could have been developed by the coal industry to compete with nuclear power in premium markets. Nevertheless, he suggests that gas conversion should be developed by the mid 1980s to enable nuclear power to provide base load electricity. By that time, the nuclear industry would have a clearer picture of uranium prices and reserves, and more operating experience of burner reactors to assess if, and when, breeder reactors should be developed.

If the coal industry diversified its market outlets by developing SNG, some of the criticisms of the Government's strategy put forward by Manners would be answered. Manners contests the NCB high investment proposals to expand production and offers a low investment strategy tailored to demand as an alternative policy.[16] In the short term he forecasts that oil and gas will dominate indigenous production with coal production falling to between 85 and 100 million tons by 1985. Economic models predict a better performance for the coal industry – 124 million tons in 1986[17] – but this figure is still below the NCB's target of 135 million tons. Although Manners argues convincingly that there is an insufficient market for this level of production, this is a short-term approach to a long-term problem. The OECD, CIA and the Workshop on Alternative Energy Strategies produced reports in 1977 emphasising the depletion of oil reserves, regardless of OPEC policy.[18] A shortage of crude oil in OECD countries could occur in the first half of the 1980s and the oil industry would need to discover the equivalent of North Sea reserves each year to avoid a world shortage by the late 1980s.[19]

Clearly, the period of an oil-based economy is nearing its end, and the three reports urge that action is taken as soon as possible to secure supplies for the future. With the nuclear industry undecided over its choice of reactor and public opinion hardening against waste reprocessing, an expansion of the coal industry is imperative for Britain's long-term future. Considering the long lead times between the planning, construction and operation of any major development, the NCB, despite its short-term problems, will require additional capacity by the 1990s when there will be fierce competition for high priced, scarce fuels on the world market.

An overall integrated energy strategy is necessary for future long-term planning but the role of the Energy Commission in policy formulation has been questioned by various bodies. NALGO's Energy Policy Advisory Committee accepts that the creation of the Commission is a step in the right direction but its role is weakened because it is only an advisor to the Government.[20] NALGO believes that a National Energy Corporation should be established on the lines of the National Enterprise Board to implement energy policy decisions after the full range of options have been discussed. The Town and Country Planning Association (TCPA) considers the Commission to be too one-sided in its representation to give an unbiased analysis of Britain's energy situation.[21] TCPA recommended that a Standing Royal Commission should be set up to propose an energy strategy within two years in order that future planning inquiries could be resolved by reference to the national strategy. The 'national interest' is usually given priority over local opposition to proposed energy projects. Hence the proposal by the Government in 1978 to create an independent high level Standing Commission which would advise on the interaction of energy policy and the environment.

Notes and references

1. In 1951 the Iranian Government nationalised the concessions of the Anglo-Iranian Oil Company but retaliation by the oil majors, which controlled the tanker fleets and the distribution networks, ensured a return to the status quo in the winter of 1953/54. The two Suez crises of 1956 and 1967 temporarily curtailed oil supplies to Western Europe.
2. E. J. Epstein (1975) *The Secret Deals of the Oil Cartel: an illustrated history*, Part II, New York, 30 June, p. 57.
3. In 1972 the US imported 30 per cent of its oil requirement: in 1976 this figure increased to 45 per cent.
4. Over one-half of the oil transported around the world in the 1970s was shipped by Very Large Crude Carriers (VLCC) of 200,000 to 300,000 tons or Ultra-Large Crude Carriers (ULCC) of over 300,000 tons. See N. Grove, 'Giants that move the world's oil: superships', *National Geographic*, **154** *(1)*, *July 1978* and N. Mostert, *Supership*, Penguin 1976.
5. Peter R. Odell. Ch. 2 in M. Saeter and I. Smart (1975) *The Political Implications of North Sea Oil and Gas*, IPC Science and Technology Press Ltd.
6. Ibid.
7. G. L. Reid (1973) Ch. 5, p. 237 in G. L. Reid, K. Allen and D. J. Harris, *The Nationalised Fuel Industries*, Heinemann.
8. *Fuel Policy*. Cmnd 3438, HMSO 1967, p. 55.
9. Cambridge Information and Research Services. *Energy Markets to 1990*, CIRS 1977, and BP Report quoted in the *Financial Times*, 10 March 1977.
10. Department of Energy (1976) *National Energy Conference 22 June 1976 vols I and II*, Energy Paper No 12, HMSO.
11. Department of Energy (1976) *Energy Research and Development in the United Kingdom*, Energy Paper No 11, HMSO.
12. Department of Energy (1977) *Energy Policy Review, Energy Paper No. 22*, HMSO.
13. R. W. Todd and C. J. N. Alty (eds). (1978) *An Alternative Strategy for the UK*.
14. Royal Commission on Environmental Pollution, Sixth Report (1976). *Nuclear Power and the Environment*, HMSO, p. 187; and Department of Energy (1977), *Energy Policy Review*, op. cit., p. 25.
15. N. Forman (1978) 'A fuel policy for posterity', *New Scientist* 16 March 1978, p. 720.
16. G. Manners (1978) 'Alternative strategies for the British coal industry', *Geographical Journal*, **144**, *Part 2, pp. 224–34.*

17. The consultants qualify this prediction in that only 105 million tons will be demanded if coal prices rise 6 per cent faster than inflation; see *Energy World*, 46, March 1978, p. 5.

18. OECD (1977) *World Energy Outlook*; Central Intelligence Agency (1977) *International Energy Situation* CIA; Workshop on Alternative Energy Strategies (1977) *Energy Global Prospects 1985–2000*, McGraw-Hill.

19. David Steel, Chairman of BP, quoted in the *Financial Times*, 24 May 1977.

20. NALGO Energy Policy Advisory Committee (1978) *A Planned Energy Policy*, NALGO.

21. Town and Country Planning Association (1978) *Annual Report*, p. 14.

Chapter 2

Coal, planning and the environment

The British coal industry has undergone significant changes throughout this century from the peak production year of 1913 to the 1970s, when in 1974 total deep-mined production fell to 98 million tons – the level of coal exports in 1913. The loss of the export market during the inter-war period and the collapse of the domestic market in the late 1950s and the 1960s caused gradual contractions of the industry, which if maintained would have resulted in the decline of production levels to less than those of the mid-nineteenth century. The 'energy crisis' during the winter of 1973/74 and the ensuing escalation of oil prices transformed the energy situation. Coal was once again cost competitive and, for the first time since the 1950s, the NCB planned an expansion of output.

In the first part of this chapter, an attempt will be made to trace the development of the industry with particular reference to the post-war period. An evaluation of the plans for coal will be undertaken to assess the basis upon which decisions were made and to monitor the success or failure of the forecasts. The remainder of the chapter will be devoted to the environmental problems associated with the extraction of deep-mined and opencast coal. In our industrial society, the increasing demands for energy and the price we are willing to pay is often reflected in our environment. The scars of dereliction from coal mining remain, often long after an area has witnessed its final pit closure. Nevertheless, we will continue to need coal and the mines of the future will in some cases be in virgin areas, previously unblemished by the spoils of extractive industries. What will be the consequences of coal developments in these areas? What becomes of colliery buildings and spoil heaps which have outlived their usefulness? These questions will be answered and illustrated with examples of the planning for new, and the reclamation of old, and redundant coalfields.

The pre-nationalisation period

Coal has been mined in Britain for thousands of years but large-scale extraction centred on present-day coalfields is a product of the Industrial Revolution. In the Middle Ages coal was worked for local consumption although the main market for coal was in the populous but coal-deficient south-east of England. Much of this demand was met by Northumberland and Durham coal which established pre-eminence over inland fields because of the marketing advantages of seaborne transport. In 1660, 2 million tons of coal were produced in Britain and over half a million tons were being shipped from Tyneside.[1] The north-east monopoly continued for over a century until improvements in inland transport –

by canal and later railway – encouraged the development of new coalfields in the first half of the nineteenth century.

Innovations during this period stimulated home consumption of coal. For example, Abraham Darby's introduction of coke smelting to the iron industry greatly increased the demand for coking coal and the new availability of cheap iron revolutionised coal machinery, which had hitherto been built of traditional materials – mainly wood and brass. The steam engine, invented by Newcomen and later refined by Watt, allowed the exploitation of deeper reserves to meet the increased demand for coal by industrial users. In the second half of the nineteenth century, the expansion of output was mainly related to the growth in export markets. The replacement of sailing vessels by steam ships stimulated demand for British coal and coal ports in South Wales, the North-East and Scotland thrived on an export trade which accounted for 12 per cent of total production in 1873 and 34 per cent in 1913.[2]

The year 1913 marks a watershed in the history of the British coal industry. The production of 287 million tons, of which 98 million tons were exported, has never again been equalled and the industry began to contract slowly in the three decades prior to nationalisation. After the First World War, freak market conditions concealed the fall in demand for coal. The coal shortages in the immediate post-war period were aggravated by the American strike of 1922 and the French invasion of the Ruhr in the following year. British coal was in demand. The remainder of the inter-war period, however, was marked by a slowly declining export demand for coal, and by as early as the late 1920s the export market had contracted to a half of its pre-war size. In Britain, however, consumption had stabilised and the depression coupled with greater efficiency in fuel burning resulted in a negligible rise in demand from the 184 million tons consumed in 1913 to 185 million tons in 1937.[3]

The collapse of export markets and the stagnation of the domestic market was partly due to factors outside the industry's control. Nevertheless, British coal became progressively less competitive in world markets. Productivity was higher in continental mines, which were newer, larger and more mechanised (Table 2.1). To increase productivity and reduce costs, new mines using

Table 2.1 Productivity in selected countries (*Source:* National Coal Board, *Plan for Coal* 1950)

Country	Basic year*	OMS† in basic year (cwt)	OMS in 1936 (cwt)	Increase (%)
Poland	1927	23.44	36.20	54
Holland	1925	16.48	35.94	118
The Ruhr	1925	18.62	33.66	81
Britain	1927	20.62	23.54	14

* The year when each country regained its pre-1914 level of productivity.
† Output per manshift.

modern equipment would have been necessary to replace the old pits which had exhausted the best seams and were working poorer, thin seams. This objective was never fully realised because of the dual problems of poor industrial relations and the fragmented structure of the industry. The large number of small, independent mining companies which determined output and prices in their districts inhibited rationalisation of the industry and were unable to provide the scale of investment required to compete in foreign markets. The continuing disputes between the management and the mineworkers did nothing to improve the position. From 1900 to 1949, 60 per cent of all man-days lost through disputes in British industry were attributed to the coal industry.[4] Even during the Second World War industrial relations were poor and labour shortages occurred because of the better wages in the munitions and other heavy industries.

The Government recognised the problems within the industry during the inter-war period and introduced measures in an attempt to improve its competitiveness. The Sankey Commission of 1919 and the Samuel Commission of 1924/25 recommended the amalgamation of companies into more economically viable units. Subsequent legislation encouraged these aims but legal disputes over mineral rights proved to be an impediment to rationalisation, whilst the quotas and fixed minimum prices embodied in the 1930 Coal Mines Act militated against amalgamations because the scheme protected inefficient pits.

The Second World War accentuated the inadequacies of the industry. Large amounts of capital investment and a rationalisation of the administrative structure to facilitate national planning were prerequisites for the future rejuvenation of coal mining in Britain. The war-time Government became increasingly involved in the running of the industry. Initially, Government money financed mechanisation schemes to improve productivity and increase output for the war effort, but the shortage of labour in 1942 provoked the creation of a state of emergency and the Government enforced its powers and took control of the industry. In 1945 the Reid Report recommended the reorganisation of local production on a coalfield basis with modifications in the system of ownership.[5] Nationalisation was inevitable. The post-war Labour Government accepted the Reid recommendations and on 1 January 1947 the National Coal Board was formed to become the largest employer in the country, inheriting 800 independent undertakings with an annual turnover of £360 million a year.[6]

1947 to 1957: the decade of expansion

The NCB could not have taken over at a more unfortunate time. The fuel shortage throughout post-war Europe was aggravated by the worst winter for many years and emphasised the need for an expansion of output. From 1947 to 1949, £72 million was invested in the industry and deep-mined output rose by 15 million tons to 203 million tons during this period. As energy demands increased at home and abroad, the NCB's main concern was to expand output at all costs. The *Plan for Coal* produced in 1950 proposed a £635 million programme of capital investment over the 15-year period to 1965 in order to raise output to 240 million tons per annum of deep-mined coal. The Ridley Committee considered that this target was too low since home consumption would increase on the phasing out of coal rationing. Their forecast of coal demand in the early 1960s was 260 million tons.[7] This report was published in 1952, the peak production year of the post-war period, and the Committee perhaps felt that output would continue to rise. Production had increased in all divisions since 1947, although most of this increase had occurred in the first two years, but from 1952 to 1957 only the low-cost East Midlands coalfield continued to expand output (Table 2.2).

Table 2.2 Coal production (million tons) by NCB Division, 1947 to 1967 (*Source: National Coal Board Report and Accounts, 1966/67*)

Division	1947	1952	1957	1962	1966/67†
Scottish	22.1	22.8	21.0	17.5	14.4
Northern and Durham*	34.7	38.5	37.4	33.8	27.2
North-Eastern	38.0	44.8	43.3	42.2	38.9
North-Western	14.3	16.6	16.3	13.2	9.7
East Midlands	35.0	43.8	46.7	46.3	42.6
West Midlands	16.6	17.9	17.4	14.3	13.5
South-Western	22.3	24.7	23.6	18.8	16.8
South-Eastern	1.4	1.7	1.7	1.5	1.5
Total (deep-mined)	184.4	210.6	207.4	187.6	164.6
Opencast	10.2	12.1	13.6	8.1	7.1
Licensed and other mines	2.2	2.1	2.7	1.7	1.3
Grand Total	196.8	224.8	223.6	197.4	173.0

* The Northern and Durham divisions were amalgamated in 1963.
† From 1963 the NCB report and accounts were produced in March instead of at the end of the calendar year.

Throughout the 1950s the Coal Board could not meet the demand for fuel. Inland energy consumption increased whilst coal production began to stabilise and as this gap widened coal exports were curtailed and diverted to the home market. The fuel shortage of 1955 was so acute that 11.5 million tons of coal were imported and the Government produced its first plan for nuclear energy to supplement coal supplies.[8] Industries were encouraged to use coal more efficiently or to use alternative fuel resources. The railways were encouraged to develop other forms of traction and the electricity industry converted power stations to oil burning.

It is somewhat paradoxical that in the inter-war period British coal production slumped because of a lack of demand but in the early post-war years, the industry could not supply buoyant markets. Why was this? The main problem was the time-lag between the injection of capital and the realisation of increased output. Despite vigorous recruitment campaigns, employment levels remained constant over the ten-year period (Table 2.3). The shortage of labour, especially skilled labour, led to further delays in the implementation of the reconstruction programme. Only 20 out of a total of 167 schemes begun in 1947 were completed in 1955.[9] This slow rate of development delayed the opening of new, more efficient collieries to boost production and extended the life of old, unproductive pits in order to maintain output levels. Indeed, the main reason for pit closures from 1947 to 1957 was the exhaustion of reserves (Table 2.3).

The combination of heavy capital investment and the continued working of inefficient pits contributed to a near doubling in the cost of coal between 1947 and 1957.[10]

The high costs and low productivity pits were located in the South-Eastern, South-Western, Scottish and Durham divisions (Table 2.4). This contrasted with the East Midlands, and to a lesser extent the Yorkshire coalfield (North-

Table 2.3 NCB employment and number of mines (*Sources: National Coal Board Report and Accounts 1966/67* and *1967/68; Digest of UK Energy Statistics* 1977)

Year*	Number of mines	Manpower (000's)
1947	958	710.5
1952	880	713.5
1957	822	703.5
1962	616	536.2
1964	576	505.3
1967	438	409.7
1969	317	318.9
1971	292	286.4
1972	289	274.0
1974	259	242.5
1975	246	248.8
1976	241	243.7
1977	238	242.1

* From 1947 to 1962, the figures relate to the situation in December of each year, from 1964 to 1977 to the situation in March of each year.

Table 2.4 Costs and productivity by NCB division, 1957 and 1967 (*Source: National Coal Board Report and Accounts, 1957* and *1966/67*)

Division	Costs/Saleable ton (£.p)		Output/Manshift (cwt)	
	1957	*1967*	*1957*	*1967*
Scottish	4.80	5.55	20.0	32.4
Northern	4.14	5.19	25.3	31.5
Durham	4.71		20.3	
North-Eastern	3.75	4.68	27.2	38.5
North-Western	4.65	6.36	21.9	29.3
East Midlands	3.08	3.89	37.9	50.9
West Midlands	3.96	4.86	25.8	40.2
South-Western	4.92	6.27	18.9	26.5
South-Eastern	5.03	5.80	21.1	30.3
Great Britain	4.08	4.92	24.9	36.6

Eastern division) where costs and productivity were consistently better than the national average. The advantages of three high cost areas – South Wales, Kent and Durham – were in their production of special quality coals, such as anthracite and coking coal (Fig. 2.1). The divisions with the lowest mining costs, on the other hand, were producers of general coals.

The NCB, aware of the stabilisation in output, revised its *Plan for Coal* in 1956.[11] The new plan confirmed the output target of 240 million tons by 1965 but modifications to the original proposals were recommended. Opencast mining which was to be phased out in the original plan, had made a significant contribution to meeting the demand for coal in the early 1950s. The Board acknowledged this and scheduled 10 million tons of opencast coal in their revised estimates. By increasing investment and expanding output from the low cost East Midlands field, the plan envisaged that the 1965 production targets could be achieved.

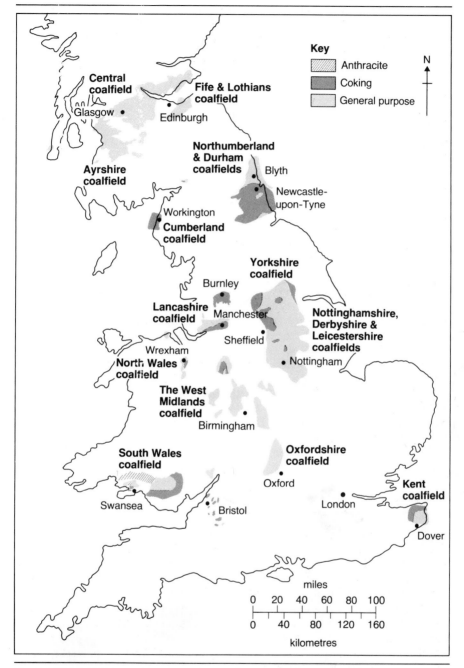

Fig. 2.1 The coalfields of Great Britain. Location of the main classes of coal (*Source:* National Coal Board, March 1979)

It is unfortunate that the decisions taken and the plans formulated from 1950 to 1956 emphasised a continued expansion of output. Neither the NCB nor the

Ridley Committee anticipated a fall in demand for coal. Consequently, the industry – planning for expansion – found it difficult to adjust to the changing market conditions of the 1960s. On the financial side, the Government had encouraged the industry to borrow to finance its capital requirements instead of securing funds through higher prices, and debts mounted as output fell.

1957 to 1967: the decade of contraction

The decade of expansion and energy shortage was followed by a period of energy surplus and the availability of cheap energy supplies. Coal could not compete and by 1967 deep-mine production had fallen by 20 per cent and the labour force was trimmed to 58 per cent of its 1957 size (Tables 2.2 and 2.3). Pit closures were a feature of the 1960s; initially, old mines nearing exhaustion were prematurely closed but by the second half of the 1960s many were axed simply on economic grounds. From 1964 to 1969 pits were closing at the rate of one per week. The motive of the 1950s to increase output at all costs was replaced by the aim of cutting costs and improving productivity. The rate of closure enabled the NCB to secure these objectives. From 1947 to 1957 the cost of coal had doubled, from 1957 to 1967 it only increased by 20 per cent; similarly productivity only increased by 14 per cent in the decade of expansion but an improvement of 47 per cent was achieved during the next decade (Table 2.4). Closures affected all divisions, including the most productive, the East Midlands division. The high cost, low productivity divisions, in particular, suffered from the NCB cuts. The coalfields most acutely affected within these divisions, such as Fife, South Wales, Cumberland, Northumberland and Durham, were granted Special Development Area status in 1967 in order to attract alternative forms of employment to replace coal mining. Unfortunately, mining had been the predominant source of employment in many towns and villages in these areas. Their economic base had been removed and their relative inaccessibility was a locational disadvantage for the creation of new industry.

The decline of the coal industry was rapid. Within three years of the NCB's endorsement of its expansion plans in 1956, undistributed coal stocks had reached their highest level in the post-war era. Internal consumption of coal fell. Initially the fall in demand was attributed to the mild winter of 1957 and the recession of 1958 but, as coal stocks accumulated, it became apparent that these trends were deep-seated rather than temporary in nature.

The surplus coal could not easily be sold abroad. Competition was keen with other European producers, and the Americans had obtained a foothold in the market as a result of contracts drawn up in the years of shortage. The NCB in its 1958 report began to plan for a decrease in output and the *Revised Plan for Coal* of 1959 downgraded previous estimates for 1965. The plan's emphasis on improved productivity and cost competitiveness by a concentration on fewer, more modern collieries resulted in a forecast of 206 million tons by 1965. The actual output for that year was 192.5 million tons. Coal was uncompetitive and, if the Government had not taken policy action in the 1960s, the rate of rundown of the industry would have become too great.

The markets for coal

The problems of the coal industry had moved in less than half a decade from trying to increase output to the search for markets. Indeed, the shortages during the early 1950s were partly responsible for the industry's problems in the 1960s. The Government had encouraged the use and development of other fuels and these began to break into markets previously dominated by coal.

The competitiveness of alternative fuel supplies is not the sole reason for the decline in the market for coal. Two of the main consumers of coal – the electricity and iron and steel industries – experienced considerable improvements in fuel burning efficiency. The larger size of power stations and generating sets increased thermal efficiency whilst improvements in blast furnace design and technology reduced coke requirements in the iron and steel industry.

Despite their technological advances in power generation, the Electricity Boards had become the dominant outlet for coal supplies by 1967 (Table 2.5). Nevertheless, more power was generated in nuclear and oil-fired stations throughout the decade. In 1957, the electricity industry used 46.5 million tons of coal and 1 million tons coal equivalent of oil; in 1967, the ratio of coal to oil had narrowed when 66.75 million tons of coal and 11 million tons coal equivalent of oil were burned. The nuclear industry was also competing with coal in the generation of electricity, producing 21.3 GWh in 1967, the equivalent of 9 million tons of coal, compared to negligible amounts in 1957.[12]

The gas industry's switch from coal to oil gasification was more spectacular and the figure of 16.05 million tons of coal consumed in 1967 marked the beginning of a massive fall in demand for coal. By 1969 the new oil gasification plants had come into production and the coal consumed in that year fell to 9.2 million tons. The late 1960s heralded the dawning of a new era for the gas industry – the conversion to natural gas – and the NCB was to lose one of its major customers for good.

Table 2.5 The markets for coal (million tons), 1957 to 1977 (*Source: National Coal Board Report and Accounts; Digest of UK Energy Statistics* 1977)

	1957	1967	1977
Power stations	46.5	66.75	77.7
Gas works	26.4	16.05	—
Coke ovens	30.7	23.8	19.3
Domestic	35.1	25.8	10.4
Collieries	7.2	3.0	1.1
Railways	11.4	1.5	0.1
Industry	37.5	21.2	9.1
Others	18.4	11.8	5.0
Total inland consumption	213.2	169.9	122.7
Exports (including bunkers	7.8	2.5	1.4
Total	221.0	172.4	124.1

Similarly, the railway market was lost, never to be regained, as British Railways implemented its modernisation plan to electrify main-line routes and phase out steam engines in favour of diesel elsewhere.

Other markets for coal declined. The slow rate of growth in the steel industry, coupled with improved fuel efficiency, resulted in coke ovens consuming 7 million tons less in 1967 than in 1957. Consumption also fell in the industrial and domestic markets. (Table 2.5). The image of coal was poor relative to its competitors, which offered cleanliness, ease of storage, adaptability and controllability of use. An additional factor which curtailed the use of coal was the pollution control measures of 1956. The Clean Air Act of that year was a Parliamentary response to the smogs of the early 1950s and embodied in the legislation was the designation of smokeless fuel zones in urban areas. More generally, the 1960s were marked by slum clearance programmes, the redevelopment of city centres and the replacement of Victorian back-to-back dwellings with modern flats and office blocks. The latter were more suitable for alternative fuels and a large part of the domestic market disappeared as the chimneys of the nineteenth century buildings were reduced to rubble.

The rapid run-down of the industry from 1957 to 1967 would have been more acute if the Government had not introduced measures to cushion its decline. The NCB had argued that its past forecasts had been approved and encouraged by the Government and that public money should be made available to support the industry. Government aid came in two forms – direct financial assistance and measures to protect the industry. Throughout the 1960s coal imports were banned, public buildings were encouraged to use solid fuel from April 1965, the rate of duty on fuel oil was increased and the CEGB slowed down its programme to construct oil-fired stations. The Coal Industry Act of 1967 stipulated that the CEGB had to burn more coal with the Government meeting the extra costs involved. In terms of financial assistance the Government wrote off £415 million capital debt to the Exchequer in 1965 and agreed under the Coal Industry Acts of 1965 and 1967 to pay a large proportion of the social costs of colliery closures.[13]

1967 to 1977: the decade of uncertainty

In the preceding two decades the fortunes of the coal industry had been largely determined by market conditions. In the 1950s energy shortages had stimulated the demand for coal whilst in the 1960s the abundance of cheap energy supplies had found coal at a competitive disadvantage. In the first half of the 1970s political forces began to influence the relative costs of coal and oil, its main competitor. Imported oil was no longer cheap, as a result of OPEC price increases in 1973/74, but the large wage settlements after crippling disputes in 1972 and 1974 had eroded much of coal's cost advantage.[14]

The Government and the Coal Board had not foreseen the 'energy crisis'; the run-down of the industry continued and the Ministry of Power in the 1967 White Paper forecast a decline in coal markets to around 120 million tons by 1975.[15] The NCB did not dispute that the industry needed to rationalise further but they claimed that a slower rate of run-down would be more manageable and a target figure of 135 million tons for 1975 would be more realistic.[16] By 1971 it appeared that the NCB forecasts for 1975 would be the more accurate of the two. Serious coal shortages developed in the winters of 1969/70 and 1970/71 and undistributed stocks fell from 24.9 million tons in March 1969 to 6.2 million tons in March 1971. The NCB confidently predicted an upturn in the market for coal and upgraded their original estimate of 135 million tons to 150 million tons by 1975. Nevertheless, although fewer mines were closing and employment levels began to stabilise by 1971, consumption and production continued to fall. (See Tables 2.3 and 2.6).

Long-term trends have been difficult to discern since 1971 because of the political events of 1972, 1973 and 1974. Production and consumption figures suffered in 1971/72 and 1973/74 as a result of the strikes in January 1972 and February 1974 and the escalation in oil prices which followed the Middle East October war in 1973. During both these periods of industrial action, a state of emergency was declared and measures were taken to cut consumption. The 1974 dispute had deep political overtones. The Arab embargo on oil together with the miners' strike, necessitated the three-day week and ultimately led to the fall of the Heath Government.

Table 2.6 Coal production and consumption, 1967 to 1977 (million tons) (*Source: Digest of UK Energy Statistics* 1977)

	Production					Consumption		
	NCB mines	Licensed mines	Opencast	Slurry recovery	Total	Inland	Exports (incl. bunkers)	Total
1966/67	164.8	1.1	7.1	1.8	174.8	169.9	2.5	172.4
1967/68	162.8	1.0	7.1	2.8	173.7	165.4	2.0	167.4
1968/69	153.1	0.9	6.6	2.6	163.2	165.0	3.1	168.1
1969/70	140.0	0.8	6.6	2.7	150.1	159.1	3.5	162.6
1970/71	133.4	0.7	8.3	2.7	145.1	148.3	3.0	151.3
1971/72	109.2	0.7	10.4	2.2	122.5	126.4	2.1	128.5
1972/73	127.0	0.7	10.5	2.3	140.5	128.1	2.2	130.3
1973/74	97.1	0.6	9.4	1.3	108.4	119.9	2.1	122.0
1974/75	115.0	0.6	9.5	1.4	126.5	125.2	2.1	127.3
1975/76	112.7	0.6	10.5	1.1	124.9	120.3	1.4	121.7
1976/77	106.8	0.6	11.5	1.4	120.3	122.7	1.4	124.1

The sharp increases in oil prices encouraged the coal industry to capitalise on its newly restored cost advantage by investing in its long-term future. The *Plan for Coal* produced by the NCB in 1974 was endorsed by the Government, who claimed that 'on a basis of commercial pricing the coal industry has now the capability for the first time for many years to bear its full production costs and still compete overall with oil'.[17] The NCB proposed to maintain deep-mined output at around 120 million tons by 1985 and to increase opencast production to 15 million tons a year.[18] In order to achieve these production targets the Board required additional capacity. The average age of a NCB mine in 1974 was 80 years and therefore 2 to 3 million tons of capacity would be lost each year through the exhaustion of the older pits. With a capital investment of £600 million, the Board hoped to augment deep-mined production by
1. Nine million tons a year from extending the life of pits that would otherwise become exhausted
2. Thirteen million tons a year from major new schemes at existing pits
3. Twenty million tons a year from new collieries.

Undoubtedly the NCB can achieve a deep-mined output of 120 million tons by 1985. Coal exploration has intensified since the publication of the Plan and in 1975/76 ten times more bore holes were drilled than the yearly average in the 1960s.[19] From July 1974 to November 1978, 500 deep bore holes were drilled and recoverable reserves were proved at the rate of 500 million tons a year – over four times the annual rate of consumption. The largest reserves have been discovered within or on the periphery of existing coalfields (see Table 2.7 and Fig. 2.2). Although the new reserves discovered near Coventry can be won from existing mines, the largest reserves are located in rural areas where planning approval will be necessary to sink new mines.[20]

* The Selby Coalfield has reserves amounting to 2,000 million tons but only the Barnsley seam is being worked. This seam contains 600 million tons of reserves but only 55 per cent extraction is planned to keep subsidence within acceptable limits.

Table 2.7 Recoverable coal reserves, 1977 (*Source: Energy World,* No. 34, January 1977; NCB Public Relations Offices)

Location	Reserves (million tons)
Selby	330*
Vale of Belvoir	510
Park (Staffordshire)	100
S. W. Coventry	100
Margam	30
Musselburgh	50
Extension of existing mines	500

The main problem for the NCB is not the proving of reserves but the extraction of coal at a competitive price. Since the 'energy crisis', coal prices have increased at a faster rate than the general level of inflation. Productivity levels fell in 1975/76 and 1976/77, and therefore coal's competitive advantage over oil has narrowed each year. Indeed, the chairman of the CEGB, presenting the 1976/77 annual report, warned the coal industry that if the NUM's claim for £135 a week was upheld and productivity did not improve, the CEGB would have no alternative but to switch to a greater use of oil.[21] It can be argued that the industry is not bearing its full production costs as envisaged by the Government in 1974 and that coal now has little or no cost advantage over oil.[22]

Industrial relations between the NUM and the NCB have soured in the 1970s during a period of hope for the industry, after years of decline. During the 1960s industrial relations were good despite the social upheavals associated with the large number of pit closures. The closure of old and uneconomic pits combined with increased mechanisation led to substantial improvements in productivity. Output per man-shift increased from 24.9 cwt in 1957 to 42.5 cwt in 1968/69. The NCB predicted in its 1968/69 report that further improvements could be achieved and 75 cwt per man-shift could be attained by 1975/76, thereby cutting the cost of coal to ensure a larger share of the energy market. In March 1976 output per man-shift was 44.78 cwt and only 8 per cent of the labour force achieved the target figure of 70 cwt or more per man-shift. Productivity fell to 43.56 cwt a year later and the competitiveness of coal will lie squarely in the hands of the miners themselves. The early retirement scheme agreed by the NUM and the Board will not enhance productivity prospects in the immediate future as skilled workers leave the industry and their replacements have to be trained.

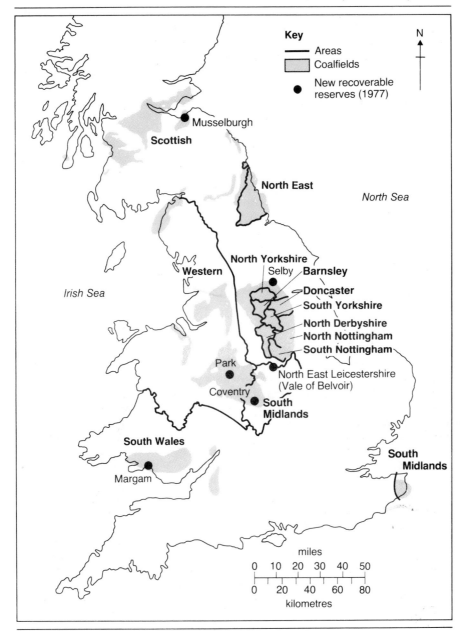

Fig. 2.2 National Coal Board areas (*Source:* National Coal Board)

The miners, like all workers, have experienced a fall in living standards as inflation rates have soared above fixed pay increases in the Government's incomes policy. Wage demands of £100 to £135 a week, however, will not be entertained by the Government or the NCB. Oil would gain a substantial competitive advantage over coal as the wage increases were reflected in higher prices, the markets for coal would decline to lower levels and the NCB would be investing in surplus capacity. To avoid a repetition of the events of the late 1950s, it is more likely that an incentive scheme will be agreed between the NUM and the NCB. This would secure the desired aim for both parties – the miners would receive more pay, and the Board would improve productivity and cut costs. An unsuccessful incentive scheme is in operation but since its introduction in March 1975, the miners have only boosted their pay packets once, a few months after its inauguration. This scheme is based on national tonnages since a pit incentive plan was voted down in a ballot the previous October. The left wing of the union believe that a pit-based scheme is a retrogressive step as it resembles the piecework system which the union successfully fought to replace by day wages in the 1960s. Although the piece-work system jeopardised safety, more coal could be won from the 722 coal faces in the country than was produced in 1976/77. Joe Gormley, the NUM president, has estimated that 8 million tons a year could be added to 1976/77 output levels if a pit-based incentive scheme could be worked out with the NCB.[23] The moderate miners' leaders, who represent the areas of highest productivity, are in favour of a scheme of this nature and detailed plans are being discussed with the NCB.

The future for coal

Negotiations between the NUM and the NCB and the periodic meetings of OPEC will strongly influence the short-term prospects of the coal industry. By 1977 the industry had lost many of its former markets and had become even more dependent on the electricity and steel industries as its main consumers (Table 2.5). Unfortunately, it is unlikely that either industry will consume more coal in the short-term. Although output has stabilised in the iron and steel industry at around 17 million tons, the industry is working under capacity. Efficiency in coke burning has continued,[24] and Warren predicts that up to 20 per cent of supplies may have to be imported by the late 1970s because of a shortage of good coking coal in Britain.[25] At present, the British Steel Corporation has signed a long-term contract with Poland to secure supplies.

Although the electricity industry has increased its consumption of coal in the 1970s, the fall in demand for electricity has made the CEGB unwilling to order the construction of the Drax B power station or to convert other stations to dual firing.[26] The CEGB claims that the added costs involved, if passed on to the consumer, would depress demand still further.[27] Furthermore, as a result of decisions taken before the oil crisis, the stations under construction are mainly oil or nuclear plants. Electricity generating capacity in Britain was 64 per cent coal-fired, 16 per cent oil-fired and 6 per cent nuclear in 1975 but the ratios will be 53, 21 and 12 respectively in 1980.[28] The NCB believes that all new fossil fuel power stations should be capable of burning coal. Until the Drax B decision, it had been over 10 years since a new coal-fired power station was commissioned. As a result, the NCB feels coal is at a commercial disadvantage compared with other fuels because of the insufficient number of modern coal-fired plants to convert coal into electricity at high efficiency.[29] The major problem in ordering more stations like Drax B is that they would merely replace the equivalent megawatts of old-fashioned coal-fired capacity. The net result of such a policy could be a fall in demand for coal.

Manners doubts the wisdom of the NCB's investment proposals because he cannot foresee a market for British coal.[30] With insufficient demand at home, he feels that the more efficient and productive pits of Poland and the USA will undercut British prices if the NCB attempts to sell surplus coal in foreign markets.

Is the 1974 *Plan for Coal* going to suffer the same fate as its 1950 predecessor? This seems not to be the case. The 1980s will not be a period of energy surplus similar to the 1960s. The 'energy gap' predicted by the forecasters mentioned in Chapter 1 will begin to appear in the 1980s as world reserves of oil and gas become depleted. Coal will be in demand in the 1980s and Britain could capture a large share of the export market. The USA, one of Britain's main competitors, has an insatiable appetite for energy and therefore, by the 1980s, America will either be directly consuming more coal at home or converting it into oil and gas for specific industries.

The lead times required between investment and output can be long through the phases of exploration, planning approval and construction. The NCB submitted a planning application to develop the Selby coalfield in August 1974; coal production will begin in 1981/82 and maximum output will be achieved in 1988.[31] Manners is wrong to assume that investment should be minimised until there is an upturn in the market for coal. The longer the delay, the more costly future developments become. The investment figures in the *Plan for Coal* have been revised from £1,400 million to £3,150 million in two years.[32]

The main long-term advantage of coal over other fuels is the vast amount of recoverable reserves in Britain – 45,000 million tons, which is over 300 years supply at current rates of extraction.[33] The NCB are not only looking to the traditional markets for coal but intend to develop technical options for the future when North Sea oil and gas resources are depleted. In 1974 the Board submitted proposals for research and development into coal conversion processes seeking Government support. The Coal Industry Examination of the same year recommended that the Government should be prepared to make a contribution in the areas of fluidised bed combustion, coal liquefaction by solvent extraction and coal pyrolysis because these projects were at an advanced stage of development.[34]

By 1977 the Government had agreed to support the most promising scheme, a collaborative project with West Germany and the USA on fluidised bed combustion. This technology offers the advantages of higher efficiencies, lower costs and a reduction in sulphur emissions in the conversion of coal to electricity. American involvement in the project was largely stimulated by the pressure of Clean Air legislation whilst increased efficiency of coal use would divert oil and natural gas to more specialised markets. Meanwhile, another coal conversion project which the Examination did not recommend for support – gasification with oxygen to yield synthesis gas – is making significant technological advances.[35] The successful operation of the slagging gasifier plant at Westfield, Fife, has encouraged the US Government to award Conoco Coal Development a $24 million contract to design a plant based on this technology.[36]

Coal and the environment

Many centuries of coal mining have left their imprint on the British countryside. Pit closures bring not only unemployment to a mining settlement, but also a legacy of years of the industry's by-products – waste tips and subsidence. Disused colliery buildings add to the scene of dereliction and hardly provide an inducement for the creation of new industry in the area. Most of the pits in operation at the present time are not subject to the planning controls embodied in post-war legislation. Nevertheless, public concern over the despoliation of the environment in recent decades has been translated into tighter planning controls on mineral workings. With the prospect of a revitalisation of the coal industry, new coal mines are inevitable but can scenes of dereliction typified by pre-1948 pits be averted in the last quarter of the twentieth century? The following sections will assess the impact of opencast and deep mining on the environment before outlining the development plans for the Selby and Belvoir coalfields. The planning implications of the Selby proposals are far-reaching, as it will inevitably be used as a test case for future applications in areas where new reserves are discovered.

DISTAND AD . OF   and Deep Mining.

Opencast mining

Since opencast mining was introduced in 1942 in order to boost war production, it has become a significant area of NCB activity. Its main advantage is flexibility of supply in that its contribution to total coal output can be dovetailed to market demand. Opencast sites, unlike deep mining sites, are temporary industrial features. Production, on average, lasts for only three or four years and the NCB contracts out the work to private tenders. Opencasting was planned to be phased out by 1945 and subsequently, in the *Plan for Coal* (1950), by 1965, but production has continued for 35 years and 341 million tons of coal have been extracted. The post-war coal shortages ensured an expansion of output, and annual production rose to an average of 11 to 12 million tons in the period from the late 1940s to the late 1950s. When demand for coal fell, the opencast section of the industry was the first to contract in order to protect employment for miners working in deep mines. Production levels languished around the 7 million tons a year mark in the 1960s, until the shortages of the early 1970s and the 'energy crisis', which gave opencasting the impetus to achieve the 1974 *Plan for Coal* target of 15 million tons by the 1980s.

Opencasting has a number of advantages. The coal is usually of good quality and can be blended with poorer deep-mined coal to meet customer specifications. Opencast production of quality coal, such as anthracite and coking coal, in itself saves the country from importing these fuels. A much higher proportion of coal to waste can be worked compared with deep mining. Of the two sectors of the NCB, deep mined coal at best breaks even on each saleable ton, compared with a profit of around £6 per ton obtained from opencast mining.[37]

The problems of opencast mining are not economic, but environmental. The Opencast Executive admits that 'an opencast coal mine is not a pretty site'.[38] The excavations become deeper each year with the use of increasingly sophisticated equipment, and draglines such as 'Big Geordie' can remove hundreds of tons of overburden in a matter of minutes. In many cases, sites have been re-opened as new technology has allowed the extraction of coal from previously unfeasible depths. Although the NCB attempts to minimise environmental disruption,[39] opposition to new sites has become more organised, resulting in protracted public inquiries.[40] If opencast's contribution to total output is to increase to 15 million tons by 1980, many new sites will be necessary, and the Government has restored the compulsory right orders embodied in the Opencast Coal Act 1958, which were allowed to expire in 1968 during the contraction of the industry.[41] Clearly the Government wishes to speed up the planning process and this is reflected in a decision made by Tony Benn, the Energy Secretary, in January 1976, to allow the extraction of 12 million tons of coal over a ten-year period at Butterwell, near Morpeth. The NCB Opencast Executive has made several applications to mine the site over a period of eight years and, after a second public inquiry, permission has been given to take 2,000 acres of dairy farmland out of production for ten years.

The Opencast Executive is well known for its high standard of restoration/ reclamation[42] and its advice is keenly sought for by planning authorities at home and abroad which wish to embark on clearance schemes for derelict land. By 1980, 133,000 acres of land will have been restored or reclaimed since opencast coal mining began in 1942.

The key to the success of the projects of the Opencast Executive is that the duration and method of mine working, in addition to the after use restoration, is planned well in advance. Coordination with the Ministry of Agriculture, Fisheries and Food is very important in this respect if the land is to return to agricultural use. The Land Service of the Ministry is involved from the prospecting stage right through until the land is relinquished by the NCB. It supervises the stripping and stacking of the topsoil and subsoil by scrapers which deposit the soil in mounds on the periphery of the site. The mounds are grassed and of sufficient height to screen the site and reduce noise levels. The overburden is usually removed by a dragline which makes a box cut to expose the coal seam. The coal is then excavated by face shovels and transported by lorries to coal preparation plants. As the dragline makes the second and subsequent parallel cuts, the overburden from each cut is deposited into the preceding void (Fig. 2.3). The overburden is regraded and contoured before the soil is replaced. The Ministry of Agriculture then assumes managerial control for a five-year period to rehabilitate the soil.[43]

The restored agricultural land will be at least of Grade 3 standard, and in many cases the Executive improves the environment of the sites that it inherits. The Pugneys site, near Wakefield, was a derelict sand and gravel quarry when opencasting began in 1972 (Plate 2.1). In 1978, after the removal of 1.3 million tons, the 205-acre site will be transformed into a country park. In Derbyshire, at Shipley Lake, the Opencast Executive was able to combine the extraction of coal with the reclamation of existing dereliction to create a much needed country park leisure area. This was opened in 1976 and includes water sports facilities, a golf course, camping area, a nature reserve and sports fields. There are many more examples of this aspect of the work of the Executive, and local authorities are beginning to appreciate the possible benefits which can be derived from a few years of opencasting, especially in areas suffering from the ravages of previous dereliction.[44]

Fig 2.1 An Open-cast Mine near Wakefield, Yorkshire

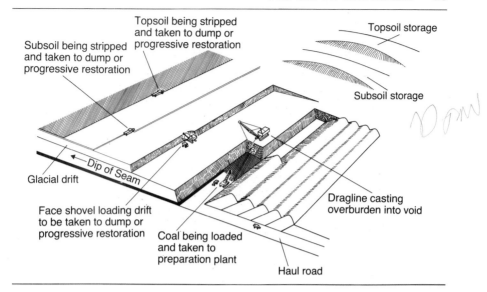

Fig. 2.3 Opencast mining (*From:* Lenihan and Fletcher: *Energy and the Environment,* Blackie 1976)

Deep mining

Many of the Opencast Executive's land improvement schemes involve the clearing up of past dereliction caused by deep mining. Whereas the Executive is obliged by law to restore the dereliction it creates, the NCB, in many instances, does not have to improve environments scarred by deep mining because most of their pits pre-date planning control legislation. The Stevens Committee collected information from fourteen counties in Britain with the object of assessing the amount of restoration work which had been achieved in mineral permissions granted since 1943.[45] Although the winning of coal by the NCB was not within its frame of reference the Committee recommended that the Board should be treated like any other mining company. In the case of private operators, the coal mining industry's record is unsatisfactory (Table 2.8). Disregarding the activities in operation, satisfactory restoration conditions had only been carried out in six cases covering an area of 43.5 acres. From this evidence, it is clear that the planning authorities are not enforcing the regulations. The Stevens Committee therefore recommended that planning authorities should strengthen their teams with mining specialists in order to monitor developments within their regions whilst, at the same time, mineral permissions should have an expiry date to make their task more manageable.

The Department of Environment's definition of derelict land has six exclusion clauses,[46] three of which enable the NCB to avoid further restoration work. These are:

(a) Land to which after treatment conditions apply.
(b) Land in current use continuing under the General Development Order 1963.
(c) Land which, although not in current use, is subject to planning permission for future development.

Table 2.8 Restoration conditions in the coal mining industry (private operators) (*Source:* The Stevens Report: *Planning Control over Mineral Working*, p. 206)

	Unsatisfactory restoration conditions	Satisfactory restoration conditions			Total
		worked and conditions complied with	worked and conditions not complied with	still working or not yet started	
Area (acres)	1,417.2	43.5	877.6	705.5	3,043.8
Number of Permissions	45½	6	7	21½	80

In the first exclusion clause (*a*) land is not classed as derelict if part of the site is still in use. Additionally, exclusion clause (*c*) allows the site to appear derelict – it is not being worked – but the Coal Board can justifiably claim that activity is only in temporary abeyance, pending new developments. The Committee felt that these loopholes were unsatisfactory and they recommended that a cessation notice should be served by the planning authority if workings were abandoned for more than six months. In the case of the NCB and dereliction caused as a result of longterm working of a mine, the Committee felt that progressive restoration should be compulsory whenever possible.

Exclusion clause (*b*), however, is mainly responsible for past dereliction attributed to the NCB. Successive General Development Orders have exempted from planning control waste tips and excavations which were in existence before July 1948 and continued to be used for the same purpose. Since the majority of pits in production are older than this, their workings are thus excluded. Before the rapid run-down of the industry in the 1960s, the proportion of older mines would have been even greater. As a result, many areas of dereliction lie outside planning control. In 1973, the NCB had 500 tips in use, 120 of which (accumulating 18 million tons of waste a year) were not subject to control.[47] At the same time the Board owned 1,500 other tips which were not used, 1,000 of which were not associated with operational collieries. On 21 October 1966 28 adults and 116 children lost their lives in the Welsh village of Aberfan in an avalanche of coal sludge from Tip Seven at Merthyr Vale Colliery – a tip not officially derelict and still in active use.[48] In response to this disaster the Mines and Quarries (Tips) Act was passed to ensure that tips would not be harmful to life and property. This Act, along with the 1956 Clean Air Act, which obliged the NCB to combat spontaneous combustion on tips, has given rise to lower, flatter tips instead of the conical and ridge types.]

Although the modern tips are easier to reclaim and can be restored progressively, they require larger areas of land than hitherto. The total area of land used for tipping in 1973 – 26,000 acres – is almost certain to increase significantly as a result of this legislation and due to the expansion programme for the coal industry in the late 1970s and 1980s. The quantity of waste deposited has even increased during recent periods of contraction in the industry.[49] The improvement of coal mining techniques has meant that by 1977 94 per cent of the material extracted was mechanically cut from the face. Prior to the introduction of mechanised mining, the miners would separate the coal from the residue at the face. The coal and residue is now transported directly by conveyor belt to the surface where coal preparation plants separate the coal from the waste.

The Stevens Committee was concerned about the production of colliery waste at a greater rate than its feasible utilisation and suggested that this waste could either be returned underground (back stowage) or that it could be separated in the mine and never brought to the surface. It is doubtful if either of these proposals could ever be implemented. The latter suggestion would be going back on mechanisation and productivity would suffer, whilst the method of back stowage has been carried out by the NCB and found to be uneconomic.[50] It is to be hoped, however, that in the future, as the NCB accumulate greater profits, it could co-ordinate its dual problems of waste disposal and subsidence. Subsidence is an inevitable problem, but it can be minimised if waste can be returned underground. At present, the NCB prefers to pay compensation for damage caused by subsidence rather than take this form of preventive action. An alternative measure is to leave pillars of coal unworked where damage is likely to be severe, as will be done in parts of the Selby coalfield.

Land reclamation schemes

Ultimately, the responsibility for reclamation of derelict landscapes created by coal mining activities lies with the planning authorities. Evidence submitted to the Stevens Committee by the County Councils shows that most planning authorities are not enforcing after use conditions attached to initial planning permissions.[51] At the same time, mines, ancillary equipment and tips which commenced operation before July 1948 are not subject to planning control and therefore the dereliction created on the abandonment of these sites is not the responsibility of the operator but of the County Council. The NCB estimates that 90 tips in this category will be discontinued by 1980, adding to a backlog of colliery wastes bequeathed to regional authorities during the 1960s.

The tackling of derelict land clearance schemes is a difficult task and the approach to this problem differs from authority to authority. The most effective approach is to set up a planning team with the full support – and financial backing – of the County Council.

However the undertaking of the vast land reclamation schemes necessary in the regions most affected by pit closures – South Wales, Scotland, Northumberland, Durham and Cumberland – required greater financial support than could possibly be forthcoming from any regional authority. In the early 1960s, most authorities would not list land reclamation as their main priority in view of their restricted and over subscribed budgets.

The Industrial Development Act of 1966 provided some of the impetus required. This gave 85 per cent grants to approved projects in Development Areas. Subsequent regional planning legislation – the Local Employment Act 1970 and the Industry Act 1972, modified the 1966 Act to give a graded scale of grant assistance. Subject to Department of Industry approval, reclamation projects can receive an 85 per cent grant in Development and Special Development Areas, 75 per cent in Intermediate, and Derelict Land Clearance Areas, and 50 per cent in other areas (Fig. 2.4).[52] From 1967 the Government sought to attract new industry to replace coal mining in the old coalfield areas. They provided extra financial incentives for firms moving to the newly created Special Development Areas, combined with grant aid. Meanwhile, other more prosperous coalfield areas were experiencing slow growth but they also had problems of dereliction which did not enhance their prospects of attracting

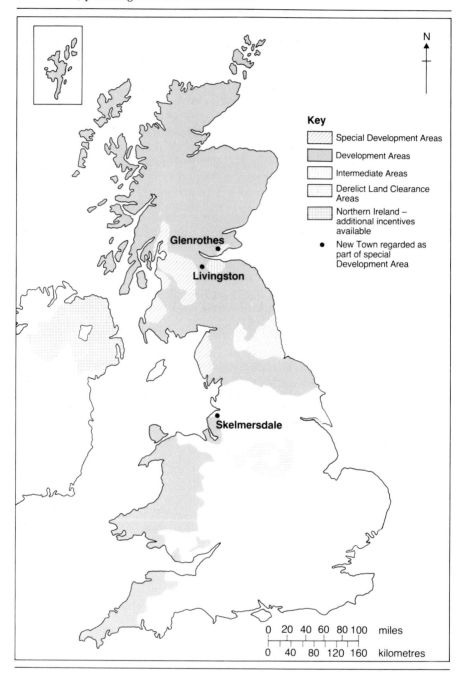

Fig. 2.4 The Assisted Areas 1972 (*Source:* Department of Industry)

industry. The Hunt Report recognised these difficulties and, by 1972, the Yorkshire and Notts/Derbyshire coalfields were designated Intermediate

Areas and the North Staffordshire coalfield become a Derelict Land Clearance Area.[53]

Although the implementation of this legislation has removed much dereliction from the British countryside, other problems associated with coal mining dereliction can take longer to eradicate.

In West and South Yorkshire, collieries have been closed but the spoils remain as adjacent collieries continue to work and others are opened or modified as part of the *Plan for Coal*. Major long-term reclamation schemes are difficult to organise in these circumstances because the NCB could still develop disused colliery sites, whilst subsidence and waste will continue to be a problem within the area. Both authorities have tended to give priority to the reclamation of sites open to public view, and tidying up operations have been carried out along the M1 and M62 motorways. For example, in West Yorkshire a new motorway industrial estate has been constructed on the former sites of the Whitwood and West Riding collieries beside the M62, and in South Yorkshire the Dodworth colliery waste tip has been re-vegetated on the side facing the M1 but left as bare shale on the other side of the tip!

The planning authorities which have been most successful in implementing land reclamation projects are those which inherited large areas of derelict land. The most frequently mentioned projects are the integrated programme of reclamation in South Wales initiated by the Lower Swansea Valley project and coordinated by the Derelict Land Unit of the Welsh Office, and the reclamation by Lancashire County Council of the abandoned coalfield area at Ashton – North Makerfield, (including the removal of the 'Wigan Alps', a series of conical waste tips) to provide 126 hectares of land.[54] In each instance, the respective authorities carried out small-scale, low cost schemes in the period up to 1966, when, now with state support, they could embark on more grandiose plans based on their previous experience. In the Welsh and Lancashire examples coal mining was mainly responsible for the dereliction, but other forms of industrial and mining waste had also ravaged the landscape in these areas.

The case of Fife

In Fife, dereliction is almost solely the product of former coal mining activity and, as the abandoned collieries were operational before 1948, the desolation was the responsibility of the County Council.[55] It is perhaps significant that Fife was the only regional authority which gave evidence to the Stevens Committee that had enforced satisfactory after use conditions on all mineral planning permissions. The region had suffered from dereliction in the past; it would not in the future.

In 1947 the NCB employed 20,600 miners at 42 pits in the coalfield. As the largest coal producing area in Scotland, the 1950 *Plan for Coal* earmarked Fife for the largest share of investment within the Scottish division in order to realise 1965 output targets. Employment in mining rose to 25,000 in 1953, but with the fall in demand for coal in the late 1950s the older pits, especially in the west, began to close. The region was dependent on coal mining – 46.3 per cent of the workforce were miners in 1959 – and the industry was contracting. In the 1950s the County was not ideally located to attract new industry because of poor accessibility. Road transport had either to cross the Forth and Tay estuaries by ferry or to cross at the lowest bridging points, Kincardine and

Plate 2.2 Lochore Meadows Phase 2, 1969 (*Source:* The Fife Regional Council)

Plate 2.3 Lochore Meadows Phase 3, 1970 (*Source:* The Fife Regional Council)

Perth. The Council was aware that the opening of the road bridges across the Forth and Tay by 1966 would create new industrial opportunities. Unfortunately, the best sites were often close to the worst areas of dereliction; for example, the projected M90 motorway from Perth to Edinburgh would be adjacent to many of the spoil heaps from the disused West Fife pits.[56] The Council appreciated the gravity of the problem and in 1959 decided to embark on a policy of land reclamation.[57] Throughout the first half of the 1960s, a series of projects, financed by the Council, were undertaken to remove not only obvious dereliction, such as pit 'bings' (tips) but also other eyesores which needed tidying up, for example, rubbish removal and grass cutting.[58]

With the advent of State aid in 1966, the planning authority began to plan the largest land reclamation project in Britain at a cost of £1 million. The Lochore Meadows site covered an area of 4 square miles and the worst scars of mining dereliction were evident in the landscape (Plates 2.2 and 2.3). At the heart of the area was a subsidence-created loch encircled by abandoned colliery buildings and miners' rows. The six disused collieries all possessed slurry ponds and tips and of the latter, three were still burning. Work began in 1967 and, on the completion of the sixth and final phase ten years later, the site had been transformed for agricultural and recreational use (Plates 2.4 and 2.5).

Most of the basic recreational facilities have been established (Fig. 2.5) and in July 1977 these attracted 1,000 people a day to a site that ten years previously had been a 'ghost' area.[59] The main recreational feature is the loch and its islands – islands created during Phase 4 of the reclamation programme. A disused mineral railway from the Mary Colliery crossed the loch and once the water level was lowered and regulated the course of the railway was severed to form a chain of islands (Plate 2.6). Another legacy of the coalfield is the Mary Colliery winding gear which has been cleaned and restored to its former site (Plate 2.5). The boats for hire on the loch also bear the names of the former collieries in the area.

Although the largest project has been completed, the programme of land reclamation will continue until the remaining scars of dereliction are removed from the County. In many ways Fife has been more fortunate than other areas suffering from dereliction. The contraction of the coal mining industry – only five mines remain in operation – has enabled the planning of comprehensive rehabilitation because sites were *officially* derelict. This contrasts with the Yorkshire coalfield where disused colliery sites might still possibly be redeveloped because coal mining still continues in the vicinity. The Fife coalfield's status as a Special Development Area from 1967 enabled the regional authority to claim the highest level of grant aid and, more recently, the Scottish Development Agency has also supported land reclamation projects. Nevertheless, it should be noted that in the period before state aid the Council financed an imaginative planning team who embarked on a programme that other authorities may have been less willing to implement.

The planning of new collieries

Land reclamation projects are only necessary because in the past coalmining was carried on without regard to its consequences for the landscape. The main objective of planning authorities in the future should be to ensure that dereliction

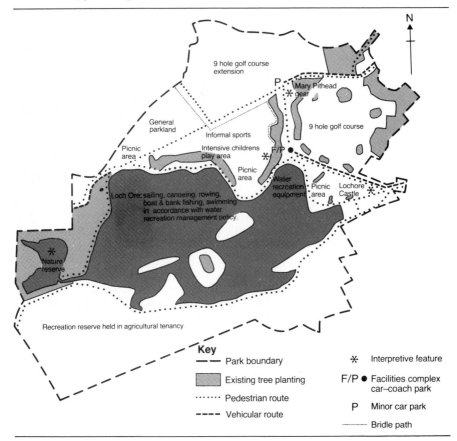

Fig. 2.5 Lochore Meadows Activities Zone Plan (*Source:* Fife Regional Council)

on the scale of 'Lochore Meadows 1967' is not allowed to occur and, if waste tipping or any other form of colliery dereliction is necessary, restoration work should be progressively undertaken instead of defacing the landscape throughout the working life of the colliery. The Department of Energy's progress report on the *Plan for Coal* envisages the possibility of 30 new coalmines being opened on established coalfields between 1985 and 2000 as well as the additional mines at Selby and the other newly discovered coalfields of the 1970s. Selby has been given planning approval; Belvoir will undoubtedly be the subject of a public inquiry in 1979. Both discoveries are in 'greenfield' areas which were planned for rural stabilisation rather than industrial expansion, and they will certainly be test cases for planning permissions in the future.

Selby

It is easy to imagine the concern of residents in the Selby area in autumn 1974 when the NCB requested planning permission to mine the Barnsley seam from within their region. The presence of unsightly collieries to the south of the area highlighted the lack of planning and the disregard for the environment that had

Plate 2.4 Lochore Meadows Phase 2, 1970 (*Source:* The Fife Regional Council)

Plate 2.5 Lochore Meadows Phase 3, 1971 (*Source:* The Fife Regional Council)

Plate 2.6 Lochore Meadows 1977 – The Loch and Islands

been associated with coal mining in the past. The Selby coalfield, however, is too rich in coal reserves – 2,000 million tons – to be left undeveloped. The Barnsley seam alone can guarantee an output of 10 million tons a year well into the next century. It has become clear that the NCB must develop the field with due regard to the quality of the environment. The Board's advertising of the project confirms this:

Old problems – new solutions
Naturally there are problems. Old problems that have been faced in mining areas for over a century. But today there are new ways of solving them. And new attitudes.
    The days of creating Blake's 'dark, Satanic mills' disfiguring the British countryside are gone. In 1976, regard for the environment is of real importance.[60]

Given this frame of reference what are the environmental problems associated with the extraction of coal from the Selby field? Much of the planning area is low lying, barely 20 feet above sea level, and the introduction of colliery buildings, in addition to the accumulation of coal spoils and stockpiles would provide a visual intrusion to an otherwise, flat, featureless landscape. Subsidence would be a major problem since it would increase the area's susceptibility to flooding, and could damage buildings of historic interest. The area is predominantly rich agricultural land and has a present population of only 30,000; a new coalmine would not only itself transform the use of vast areas of farming land but further areas would be absorbed as schools and other infrastructural services were provided for the incoming mining community. Existing communication networks would be unable to cope with the increased volume of traffic generated as a result of coal mining developments. At present the flow of road transport within the area is hampered because of the large number of railway level crossings and the bottlenecks created at towns such as Cawood and Sherburn in Elmet where routes converge.

    After a public inquiry which lasted from April to June 1975, the Secretary of State for the Environment granted permission on 31 March 1976 to mine coal, subject to certain conditions which seemed to allay many of the fears expressed about the project.

The solutions

The NCB has clearly attempted to minimise environmental disruption within the area and appointed consultants to design and supervise the structures and landscaping. If anything, the proposals by the consultants have aroused greater hostility because the shaft sites will now require double the acreage than originally intended to allow for screening, tree planting and other landscaping

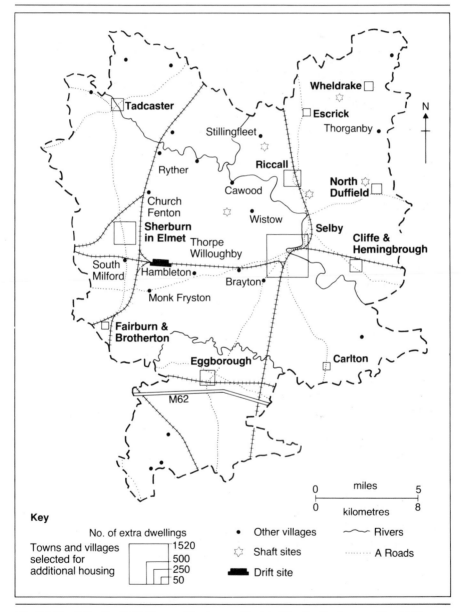

Fig. 2.6 The Selby Coalfield (*Source:* Hand-out at Regional Studies Conference, May 1977)

features. Haywood is critical of the NCB's attempt to conceal buildings which, with creativity, could be designed to blend into the landscape.[61] Concealment, he argues, is an admission that NCB designs are not fit to be seen!

Overall, however, the NCB has striven to meet the specifications laid down by North Yorkshire County Council. A pair of drift mines with a single exit point is being constructed at Gascoigne Wood, a disused marshalling yard between the villages of South Milford and Hambleton (Fig. 2.6). The total output from the coalfield will be brought to the surface at this point and the coal will then be transported *by rail* in a 'merry-go-round' system to nearby power stations. The five pairs of shafts will be used solely to convey men and materials to the face and their sites are not in areas of good quality agricultural land. Subsidence is to be limited to 0.99 metres throughout the area and pillars of coal will be left unworked to support industrial and historic buildings, especially in Selby and Cawood. The technique of controlled subsidence – whereby the area of coal worked out becomes progressively narrower and pillars more dominant – will be adopted in low-lying river basins. General agreement was reached among the parties concerned that a new mining village would be unnecessary because about half of the 4,000 miners required could be recruited locally, and the remainder could be absorbed within existing communities. Selby and Sherburn in Elmet would be the main recipient settlements (Fig. 2.6).

An extensive list of conditions is attached to the planning consent to extract coal from the Barnsley seam. Most of these have been outlined above, but other conditions include the removal of all waste material from the application area and the washing of all materials and wheels of vehicles on leaving the colliery, the restriction of the height of stockpiles to 12 metres, and a ban on the manufacture of coal by-products. The conditions, *if enforced* could make the Selby Coalfield a model example of modern coalfield planning.

Belvoir (North-East Leicestershire)

The NCB will face even greater opposition to its proposals to mine coal at Belvoir than at Selby. Although both coalfields are in rural vales, Selby is fortunate in that its underlying geology will allow the extraction of coal through one large drift mine; and one of the main environmental objections to coal mining – coal waste tips – will not be evident at Selby because of the thickness and cleanness of the coal. In Belvoir the presence of thick water bearing strata at depth precluded a drift mine on the grounds of safety and consequently the Board has proposed the winning of coal from shaft mines at three sites near Hose, Asfordby and Saltby (Fig. 2.7). Ton for ton, Selby coal is three times cleaner than Belvoir coal and, on average, the ratio of shale to coal is 1:3 in the Belvoir field.[62] This presents the NCB with a major waste disposal problem as compared with Selby. The Board proposes to combine local tipping with progressive restoration and, where appropriate, to use the waste for landscaping purposes to screen the colliery sites. The chosen sites are centred on industrial areas or close to disused airfields. The NCB stresses that these sites are not ideal mining sites but a locational compromise to minimise environmental disruption. The coal will be transported by rail rather than road and the mines have therefore been sited beside adjacent existing main lines, as in the case of Asfordby, or disused railway lines which will be reinstated (Fig. 2.7).

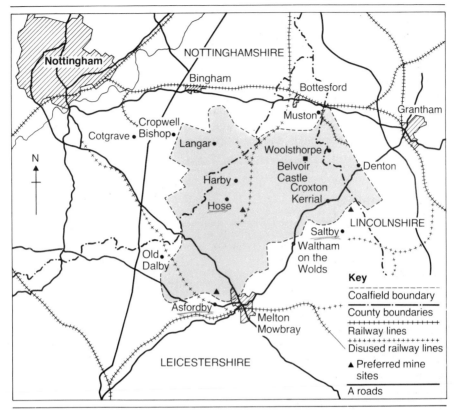

Fig. 2.7 NCB proposals for North-East Leicestershire (Vale of Belvoir) (*Source:* National Coal Board)

The three mines will employ 3,800 miners who, the NCB estimates, will require 3,500 houses.[63] It is hoped that the newcomers will be absorbed within existing communities in the same way as in the Selby area; however, a greater proportion of miners will move from other areas to Belvoir compared with Selby, where one half of the labour force is planned to come from the Yorkshire coalfield.

Although the coalfield will make a substantial contribution to the objectives of NCB's *Plan for Coal*, the environmental problems are more difficult to overcome than in Selby. The winding towers will be between 146 and 191 feet high, twice the height of the shaft head gear at Selby; the tipping of waste will be unavoidable; three shaft mines are required instead of one drift mine (albeit with five satellite shafts); the greater number of immigrant workers required at Belvoir will obviously bring social problems. All of these factors provide considerable ammunition for residents' groups in their attempts to block the NCB proposals. The Duke of Rutland, whose castle lies in a stretch of hunting country within the area of reserves, feels that the NCB has not proved to the County Council, of which he is chairman, the nation's need for Belvoir coal. There is a prospect, therefore, of a protracted public inquiry or even a planning inquiry commission, which would put the NCB's plans further behind schedule.

The first two major coalfields discovered since the NCB began to implement its 1974 proposals have been in similar locations – rural areas on the periphery of existing coalfields. Nevertheless, as our description has shown, the environmental problems associated with coal extraction vary considerably from field to field, and each planning application has to be examined on its own merits in the light of specific local conditions.

Clearly, the NCB would not have experienced so much difficulty or delay in obtaining planning approval in well-established industrial areas. It is to be hoped, however, that the Board will observe the conditions attached to the planning permissions though Haywood questions its ability to finance restoration projects.[64]

Notes and references

1. B. Lewis (1971) *Coal Mining in the Eighteenth and Nineteenth Centuries*, Longman, pp. 8, 27.
2. National Coal Board (1950) *Plan for Coal*, NCB, p. 13.
3. G. L. Reid, K. Allen and D. J. Harris (1973) *The Nationalised Fuel Industries*, Heinemann, p. 10.
4. National Coal Board (1950) op. cit. p. 15.
5. *Coal Mining. Report of the Technical Advisory Committee (Reid Report)*. Cmnd 6610, HMSO, 1945.
6. Reid, Allen and Harris (1973) op. cit., p. 14.
7. *National Policy for the Use of Fuel and Power Resources (Ridley Report)*. Cmnd 8647, HMSO, 1952.
8. *A Programme of Nuclear Power*. Cmnd 9389, HMSO, 1955.
9. Reid, Allen and Harris (1973) op. cit., p. 17.
10. *NCB Report and Accounts, 1957*, p. 1 – a summary of statistical information on the Board's progress, 1947–57.
11. National Coal Board (1956) *Investing in Coal*, NCB.
12. G. L. Reid and K. Allen (1970) *Nationalised Industries*, Penguin, p. 87.
13. For a more detailed account of the NCB Finances see *The Finances of the Coal Industry*, Cmnd 2805, HMSO, 1965.
14. Although manpower had been drastically reduced in the 1960s, wages still accounted for 50 per cent of total operating costs in 1974.
15. *Fuel Policy*. Cmnd 3438, HMSO, 1967.
16. *Select Committee on Nationalised Industries. National Coal Board*, Vol. II, p. 262.
17. Department of Energy (1974) *Coal Industry Examination, Interim Report*, Department of Energy, p. 6.
18. National Coal Board (1974) *Plan for Coal*, NCB.
19. Department of Energy (1977) *Coal for the Future*, Department of Energy, p. 12.
20. Planning approval has taken longer than expected – in the case of Selby, 19 months – and this has caused the NCB to downgrade its estimates of production from new mines in 1985 to 10 million tons instead of 20 million tons.
21. *Financial Times*, 29 July 1977.
22. G. Manners (1976) suggests this in 'The changing energy situation in Britain', *Geography*, **61**, p. 226, and the Department of Energy (1977) *Coal for the Future* p. 8 gives a breakdown of Government grants to the industry in 1974/75 and 1975/76.
23. *Financial Times*, 23 May 1977.
24. In 1965, the coke:pig iron ratio was 0.68; in 1975 it fell to 0.61. See Manners (1976) op. cit., p. 224.
25. K. Warren (1976) 'British Steel: the problems of rebuilding an old industrial structure', *Geography*, **61** p. 4.
26. The Government commissioned the construction of Drax B in April 1977 against the wishes of the CEGB.
27. See Department of Energy (1976) *National Energy Conference 22 June 1976*, Department of Energy, Vol. II, p. 33.
28. Manners (1976) op. cit., pp. 226, 227.

29. Department of Energy (1976) ibid. p. 24.

30. G. Manners (1977) *Alternative strategies for the National Coal Board* a paper presented at a conference entitled 'Energy and the Environment', Regional Studies Association, May 1977.

31. National Coal Board Press Release, 29 October 1976.

32. Department of Energy (1977) op. cit., p. 11.

33. Ibid., p. 19.

34. Department of Energy (1974) *Coal Industry Examination, Final Report*, Department of Energy, pp. 19–23, 30–6.

35. The reason for the NCB's lack of support in coal gasification schemes is that British Gas has developed this technology since the early 1960s.

36. *Financial Times*, 31 May 1977.

37. See the profit and loss accounts in the *NCB Annual Report and Accounts 1975/76* and *1976/77*.

38. Advertising feature in the *Financial Times*, 4 February 1976.

39. Preventative measures include screening the site, the use of silencers on mechanical equipment, controlled blasting and dust minimisation.

40. In July 1977 the public inquiry into the extension of the Temple Newsam site, near Leeds, was adjourned because of over-enthusiastic objections on the part of some members of the public.

41. The average number of sites in operation in 1976/77 was 56; to meet the planned production figure of 15 million tons by 1980, the number of working sites will have to increase to an annual average of 70 to 80.

42. 'Restoration' describes the process of returning a site to its previous use on the completion of coal extraction. 'Reclamation' is the process of reclaiming derelict land. It has been possible to combine both processes on some sites.

43. See the Opencast Executive leaflet 'Opencast Operations' for details of the five-year restoration programme.

44. See NCB *Annual Report 1971/72*, pp. 11, 12, and Opencast Executive publicity leaflets *Opencast Coal Mining* and *Opencast Operations*.

45. The Committee was appointed in August 1972 under the chairmanship of Sir Roger Stevens to examine planning control over mineral workings. It produced its report in March 1975 and HMSO published it in 1976.

46. Stevens Report, op. cit., p. 112.

47. Stevens Report, op. cit., p. 420.

48. J. Barr (1970) *Derelict Britain*, Penguin, pp. 11, 40.

49. K. L. Wallwork (1974) *Derelict Land*, David and Charles, p. 125.

50. Ibid., pp. 127–30.

51. For a detailed breakdown of the survey carried out by the Committee see pp. 203–351 in the report.

52. The areas qualifying for assistance have been modified considerably from 1972 to 1977 as general unemployment has led to the re-grading of assisted areas. The position in 1972 was more reflective of coal mining, rather than general unemployment (parts of N. Wales and Merseyside have been also designated as Special Development Areas in the 1970s). See Department of Industry *Incentives for Industry*, 1978, p. 4, for an up-dated map of the assisted areas.

53. *The Intermediate Areas,* (the Hunt Report). Cmnd 3998, HMSO, 1969.

54. Detailed case studies can be found in Wallwork (1974) op. cit., pp. 234–89; and Barr (1970) op. cit., pp. 79–180.

55. With the exception of the ill-fated Rothes Colliery, which began production in 1957 and closed in 1962.

56. Between 1957 and 1967, twenty-four pits closed, seventeen in West Fife.

57. Fife County Council. *Retrospect*, Fife County Council, 1959.

58. See J. McNeil (1973) 'The Fife coal industry 1947–1967: Part II' *Scottish Geographical Magazine*, **89** (3), p. 170 for a detailed breakdown of individual schemes.

59. *The Courier and Advertiser*, 20 July 1977.

60. In the *Financial Times*, 15 September 1976, and the *Guardian*, 29 October 1976.

61. I. Haywood (1977) *Energy policy: Coal, planning and the environment*, a paper presented at a conference entitled 'Energy and the Environment', Regional Studies Association, May 1977.

62. NCB press release, 18 July 1977.

63. Ibid.

64. Haywood (1977) op. cit.

Chapter 3

North Sea oil and gas

In May 1978 oil production from the North Sea reached one million barrels a day; by 1980 production will have doubled and Britain will have moved from sixteenth oil producer in the world to within the top ten. Britain, with its large population, balance of payments difficulties and ailing industrial base, will never achieve wealth on the scale of Saudi Arabia. However, Government revenues which will accrue through royalty payments and taxation could restore the country to economic health in the 1980s. While economists debate the possible strategies open to the Government, a great deal will depend on the rate of resource development. Will production peak in the mid 1980s and the wealth from offshore hydrocarbons therefore be a short-term bonanza? Much is open to conjecture. The multiplicity of factors involved makes a wide margin of error inevitable in any attempt to sketch a production profile for North Sea oil and gas. The relationship between the oil companies and the Government will determine the pace of development. The marriage of aims – the oil companies' maximisation of profits and the Government's maximisation of revenue – will ultimately decide the level and duration of production. As the oil industry moves further into unexplored waters with a more hostile environment which pushes technology to its limits, the cost of exploiting these marginal areas begins to escalate. The Government, on the other hand, can encourage or discourage the development of 'marginal' fields according to the flexibility of its fiscal policies, whilst its depletion policy can regulate production levels.

Unforeseen events can inhibit the rate of development. Current offshore installations in northern waters are the product of 'first generation' technology. Oil companies have consistently underestimated the problems of the North Sea environment, hence it is possible that relatively new platforms could be taken out of production sooner than anticipated for maintenance. The risk of a 'blow-out' cannot be discounted since the 'Bravo' incident on the Ekofisk field in April 1977. If something similar occurred again and the wells ignited in a major British field such as Brent or Forties, the cost in lost production would be measured in billions of pounds. The price paid would not only be reflected in lost production but also in the damage to the marine environment. The increased level of activity in the North Sea increases the probability of a major oil spill from tanker accidents or blow-outs. The recurrent tanker accidents occurring around the British coast do not augur well for a continued pollution-free marine environment at Sullom Voe, the main transhipment terminal for North Sea oil production.

The delimitation of offshore areas

Since the Second World War, technological developments in marine engineering have enabled oil companies to explore for oil and gas offshore. The initial exploration and development of offshore resources was in areas adjacent to traditional onshore fields, such as the Gulf of Mexico and the Gulf of Maracaibo. In both cases, the sovereignty of the area offshore was not in dispute. Nevertheless the United Nations, foreseeing potential international friction over offshore sovereign rights, convened a conference in Geneva in 1958 to formalise guidelines concerning the partition of continental shelf areas. These guidelines became international law in 1964 and a series of bilateral arrangements were made between the UK and her North Sea neighbours between 1965 and 1971. The territorial arrangements have favoured Norway and the UK, which together have jurisdiction over 71 per cent of the North Sea.[1] The median line between coastal states was drawn using the principle of equidistance. Only in the case of Germany was this principle disputed, and after the case had gone to arbitration at the International Court, the Danes and the Dutch had to concede some of their sectors to Germany in 1972. The German case was based on length of coastline rather than configuration, which gave Germany a proportionally smaller share of the offshore area. Nevertheless, despite their success, political wrangling has retarded the German exploration programme by five years. This example appears to endorse Smart's claim that the Geneva Convention was a retrogressive step in that the North Sea should have been developed as one unit with joint investment programmes and a common policy towards issues such as pollution of the sea.[2]

The German case and the more recent agreement between Britain and France in 1977 concerning the boundary in the western approaches to the Channel provide precedents for future boundary controversies. After a thirteen-year dispute between Britain and France the Channel Islands issue was resolved by granting the islands a special twelve-mile limit, thus giving the French more territory than a strict interpretation of the Geneva rules would have permitted.

Before the Franco-British agreement John Grant had speculated about the legal implications for offshore boundaries should Scotland gain independence.[3] He suggested that if Scotland became independent the SNP would claim all coastal waters north of latitude 55° 55′ N. The choice of this latitude is based on its use in the Continental Shelf Order of 1968 to distinguish between waters subject to Scots rather than English law. Such a claim would give Scotland all the oilfields and a greater share of the North Sea than any other country – 160,000 square kilometres compared with Norway's 131,000 and England's 84,000.[4] However, if Scotland did become independent, Shetland, and possibly Orkney, might either also seek autonomy or retain their existing status. In these circumstances, by the principle of equidistance, Scotland would lose all the major discoveries in the East Shetland Basin (see Figs 3.2 and 3.3) to Shetland, fields further south, including Brae and Beryl, to Orkney, and Auk, Josephine, Argyll and Fulmar to England. This would leave only the cluster of fields from Piper and Claymore in the north to Montrose and Lomond in the south under Scottish jurisdiction.

The Channel Islands ruling, however, could now significantly alter the territorial division of the North Sea. If, after Scottish separation, Shetland decided to remain within the UK, the delimitations suggested by John Grant might now be open to question. If Shetland, as part of the UK, had only a twelve-mile territorial zone, it would own no oil and could only enjoy the negotiated benefits which would accrue from the landing of the oil at Sullom Voe. Shetland could not be used as a base point in delimiting the boundary between Norway and Scotland. The new median line would be drawn using the northernmost point of Scotland as a reference point. The net result would be a shift of the line to the west, placing some northern and eastern fields (Murchison, Thistle, Magnus and Ninian) in the Norwegian sector. It is unlikely that Scotland will separate from England, but even under the *status quo* the drawing of median lines has caused problems for the efficient exploitation of North Sea reserves. Most of the sedimentary basins which contain oil and gas reservoirs straddle the boundary lines between national zones. For example, many major discoveries in the Norwegian zone lie close to or abut on to other sectors: the Ekofisk group is adjacent to the German, Dutch, Danish and UK zones, whilst further north two of the largest offshore fields – Statfjord and Frigg – are being jointly developed by Britain and Norway.

North Sea oil and gas reserves

It is important for the Government in its efforts to plan a long-term energy strategy to forecast the likely life-span of North Sea oil and gas reserves. Estimates must be revised constantly in the light of new information; many of the earlier forecasts were based only on intelligent guesswork. Until the sedimentary basins surrounding the British Isles have been thoroughly explored, estimates will reflect the limitations of existing knowledge. Nevertheless, as North Sea gas production builds up and oil moves from the development to the production stage, a clearer picture is emerging.

The main concern of the British Gas Corporation has been to tailor gas production to the increased demand caused by the rapid conversion of consumers onshore to natural gas. This has been successfully achieved with the coming on stream of the Frigg gas field, which compensates for a stabilisation of production from the older southern fields. The maturity of the offshore gas industry enables an accurate estimation of reserves to be made. The Department of Energy classifies reserves into three groups: *proven* where fields are almost certain to be developed, *probable* where fields have a greater than fifty per cent chance of being developed, and *possible*, less than fifty per cent chance of exploitation. In the southern North Sea 472 billion cubic metres (bcm) are either under development or are likely to be developed. Similarly, in the northern North Sea, 177 billion cubic metres are under contract to be exploited in the Frigg and Brent fields, (Table 3.1). At current consumption rates (40 bcm per annum) the reserves under contract would last for sixteen years. This is, however, a purely hypothetical figure. As prices increase, the remaining known discoveries (amounting to 175 bcm in the southern North Sea) are almost certain to be developed. Throughout the 1970's exploration activity in this area has dwindled with a corresponding increase in activity in the northern waters. In addition to the major finds at Frigg and Brent, the Liverpool Bay discovery extends the geographical distribution of gas discoveries in the UK sector.

Table 3.1 United Kingdom continental shelf gas reserves (Dec. 1977) (billion cubic metres) (*Source:* Department of Energy, *Development of the Oil and Gas Resources of the United Kingdom* 1978)

	Proven	Probable	Possible	Total
Southern basin				
Fields under contract to British Gas	421	14	25	460
Other discoveries believed commercial but not under contract	51	65	—	116
Other discoveries	—	31	40	71
Total	472	110	65	647
Other areas (including the Northern Basin)				
Fields under contract (Brent and Frigg)	177	—	—	177
Other significant finds (including Liverpool Bay)	40	194	314	548
Other gas, associated with oil				
1. Fields in production, under development or active appraisal	55	50	10	115
2. Other possible developments	—	—	59	59
Total	272	244	383	899
Total (UK continental shelf)	744	354	448	1,546

More gas obviously remains to be discovered in the northern North Sea. However, it is difficult to make an accurate assessment of the potential gas reserves of northern basins because gas discoveries in the north tend to be linked to oil exploration and development. Potential oil reserves can be calculated but the oil:gas ratio remains uncertain until exploration drilling is undertaken.

Additional reserves can be exploited if a gas gathering pipeline network is constructed to tap marginal gas finds not at present developed and the small deposits associated with oil discoveries which are often wastefully flared at the drillhead. Stricter control by the Government on gas flaring and its insistence on gas re-injection equipment on larger fields will lengthen the life of reserves. Overall, the continental shelf should yield adequate supplies of gas until the first quarter of the next century, and the Department of Energy estimates that known reserves are 1546 bcm with an ultimate potential of 2270 bcm.[5]

It is easier to estimate oil reserves than gas reserves, but the regulation of supply to demand is easier for gas because the British Gas Corporation is the only buyer. The British National Oil Corporation's agreements with oil companies should, however, give it access to over half of UK output in the 1980s, and a much clearer picture of reserves and peak production levels will be obtained.[6] The discrepancy in estimates varies widely, from the conservative figures produced by the oil companies to the optimistic forecasts of Professor Peter Odell. Government assessments of likely reserves are slightly more optimistic than those of the oil companies (Table 3.2). For the whole of the North Sea Odell's forecast of 79 to 138 billion barrels is more than double the estimates given by Shell and BP, which range from 50 to 55 billion barrels.[7] The British, Norwegian and Danish Governments in total estimate a range between 50 and 76 billion barrels. In the UK sector, BP predicts that ultimate recoverable reserves will approximate to 3 billion tonnes (22 billion barrels),[8] whereas the Government's present forecast is 3 to 4.5 billion tonnes (22 to 33 billion barrels).

Table 3.2 Oil reserves in UK licensed areas (million tonnes) (*Source:* Department of Energy, *Development of the Oil/Gas Resources: United Kingdom* 1978)

	Proven	Probable	Possible	Total
Fields in production or under development	1,100*	145	185	1,430
Other significant discoveries not yet fully appraised	305	480	405	1,190
Total (present discoveries)	1,405	625	590	2,620
Expected discoveries present licences (including 5th round)	—	—	600	600
Total	1,405	625	1,190	3,220

* 50 million tonnes of which have been produced.

Odell and Rosing have accused the oil companies of 'creaming' fields by only partly developing discoveries to ensure an acceptable return on investment (25 to 35 per cent after tax).[9] The gist of their argument is that it is in the interest of both the Government, which hopes to maximise revenue, and the oil companies, which desire maximum profitability, to agree to maximise output. In their study of the Forties, Piper and Montrose fields they argued that optimum production in the national interest was 3.5 billion barrels instead of the 2.76 billion barrels quoted by the oil industry. To achieve the higher figure, they assume that a greater number of production platforms will yield much greater production; for example, in the Forties field one platform would yield 708 million barrels but two platforms 2,538 million barrels. Odell and Rosing in an earlier paper used a simulation model to forecast development in the North Sea. Their model incorporated a measure of the probability of finding hydro-carbon reservoirs, the number of exploration wells drilled per licence and an 'appreciation' factor.[10] The latter factor predicts the probable upgrading of reserves from initial estimates. From past experience in other oil provinces, recoverable reserves on depletion have been four times more fruitful than original oil company forecasts.[11]

Odell and Rosing's conclusions have been severely criticised both by the oil companies and by academics. Chapman and Wall have focused their criticism on the arguable geological data used. Chapman maintains that accurate estimates cannot be made from the data available and that Odell as an 'outsider' to the industry would not have had access to confidential information from which to draw more definitive conclusions.[12] The appointment of Odell as a consultant to the Department of Energy has therefore caused the oil companies some concern about the possible release of field development plans although this would not occur without the operators' approval. Wall has accused Odell of having insufficient experience in petroleum engineering to interpret accurately the net volume of rock containing oil, porosity (the void space in the rock) and the fractions of porosity occupied by oil.[13] By recalculating Odell's figures, Wall's team estimates that recovery values would be 19 per cent lower, producing a figure of 2,145 million barrels well below even the industry's forecast for the three fields. Kemp disputes the claim that recovery levels would increase significantly if more platforms were added to a field. He suggests that decisions are taken on matters of this nature only after experience is gained during the development process.[14] He quotes in evidence the belated decision by Chevron to install a third platform in the Ninian field.

One of the issues of conflict between Odell and the oil companies over the estimation of reserves is the level of comparability between oil provinces. Odell has argued that only a fraction of exploratory wells have been drilled in parts of the North Sea compared with similar-sized areas in the Gulf of Mexico.[15] Consequently he questions both the oil companies' thoroughness of exploration and their pessimistic reserves forecasts. The industry, on the other hand, argues that no two areas are similar and that the North Sea is a virgin area in which the oil companies can only gain experience as they develop fields.

Who is right in this dispute? Odell's forecasts cannot be dismissed because BP and Shell have upgraded their estimates of North Sea reserves in recent years and historically this has proved to be the case in other oil provinces. Oil companies originally made estimates prior to the publication of the Government's taxation and participation policies. Consequently, it was in their financial interest for their original estimates to be conservative; also, the high capital investment involved in the development of the North Sea would inevitably lead to their adoption of a cautious approach. On the other hand, if much investment is to be tied up in a high-cost oil area, the companies must thoroughly appraise the likely recoverable reserves before committing themselves to a development programme. Similarly, the Government insists on a work programme being submitted prior to the issuing of licences, and the development plans for all fields have to be approved by the Department of Energy.

The increasing involvement of the BNOC will further ensure that the Government has an indication of oil company practice. Because of Government involvement and the capital investment required for field development, MacKay assumes that the 'appreciation factor' will not be as high as four, but perhaps one and half times original estimations.[16]

Ultimately, the answer will depend on the number of discoveries which have yet to be made. In other oil provinces, the main oil reservoirs were discovered first and the discovery of small to medium-sized fields were the result of a second phase of exploration and development. This pattern seems to be being repeated in the North Sea. At Groningen the successful gas discovery contained in a reservoir of Rotleigendes sandstone trapped by impermeable Zechstein salt promoted exploration in the Anglo-Dutch basin where hydrocarbons were discovered under similar geological conditions. After the Ekofisk and Montrose dicoveries, exploration intensified in the northern basins – especially in the Forth Approaches basin and the narrow Viking Graben basin (Fig. 3.1). Most of the troughs in the northern basins have been overlain with younger rocks: jurassic sands often provide the reservoir and jurassic shales such as Kim-

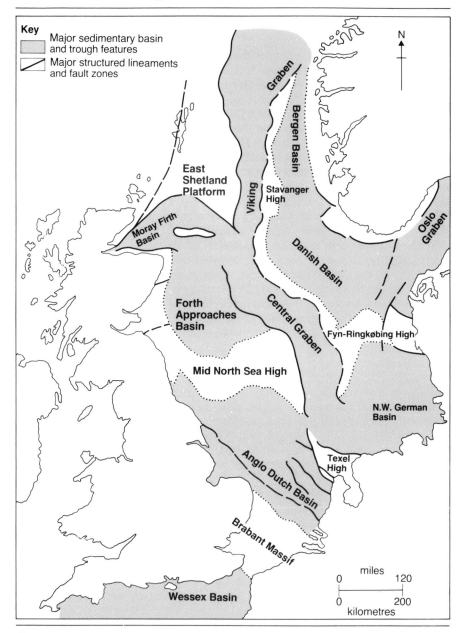

Fig. 3.1 Major mesozoic and tertiary structural units in the North Sea (*Source:* After P. E. Kent, *Journal of the Geological Society,* September 1975)

meridgian or lower cretaceous shales act as cap rocks. This is not strictly true for all fields; for example, the Forties field is located in rocks of younger age – lower eocene and palaeocene rocks.[17] Although the oil companies conduct precise seismic investigation into the basins which are liable to be the most productive before applying for exploration licenses, their knowledge is not complete. Two of the largest oil discoveries in the North Sea, at Ekofisk and Forties, were located in areas which were originally considered unattractive. Phillips was the only applicant for the Ekofisk block in 1965, and BP discovered Forties twelve months before its obligation to relinquish the block for re-allocation.[18] The oil industry is consistently surprised by events in the North Sea. Mesa discovered the Beatrice field in 1976 in the Moray Firth basin close to the shore – an area which had been unsuccessfully explored in the late 1960s.

Despite its promising discovery in block 206/8 in the West Shetland basin and the onshore discovery of Wytch Farm in Dorset, BP is not optimistic that the relatively unexplored areas around Britain will yield significant supplies of oil and gas. On the basis of the company's estimate of total reserves, between two-thirds and three-quarters of the oil has already been found. Nevertheless, most estimates will remain very vague until a more thorough exploration effort has been made in these virgin areas. BNOC estimates that 40 per cent of all wells drilled by 1981 could be in territory unexplored by the oil companies, with over 10 per cent being drilled in the West Shetland basin.[19] A clearer picture of available reserves will obviously develop in the 1980s; then, much will depend on the price of oil, the cost of extraction and Government policy to determine the form of production profiles in the 1990s.

North Sea gas

Since the discovery of the West Sole field in 1964, the role of natural gas in UK primary energy consumption has undergone a meteoric rise from negligible amounts in the late 1960s to 18 per cent of the total in 1977. The initial discovery prompted intensive exploration in the southern North Sea until 1970. By then the four major fields – Leman, one of the largest offshore gas-fields in the world, Indefatigable, Hewitt and Viking – had been located and were in production or in the process of development (Fig. 3.2).

Throughout the 1970s exploration has waned in the southern basins and attention has moved progressively further north (Table 3.3). The major reservoirs were found quickly in the south and the oil companies, which were obliged to sell their newly discovered gas to the Gas Council, looked to the northern basins for oil, which did not have to be sold to a single buyer.

The initial contracts between the Gas Council and the oil companies were almost at a 'cost' price. The Council's aims were to encourage the companies sufficiently to continue exploration, enable them to make a 'reasonable' return on investment and to keep prices to the consumer competitive enough to optimise the use of available reserves. To ensure that the gas supplies coming ashore at Bacton, Theddlethorpe and Easington (a forecast 4,000 million cubic feet per day by 1975) would match demand, the Council had to undertake a large-scale programme to convert customer's appliances from manufactured gas to natural gas.[20] By 1977, 13 million domestic gas users, 400,000 commercial users and 57,000 industrial consumers had had 35 million appliances converted from town to natural gas.[21]

Having successfully promoted 'high-speed gas' and finalised the conversion programme, British Gas has had to secure supplies for the longer term. With

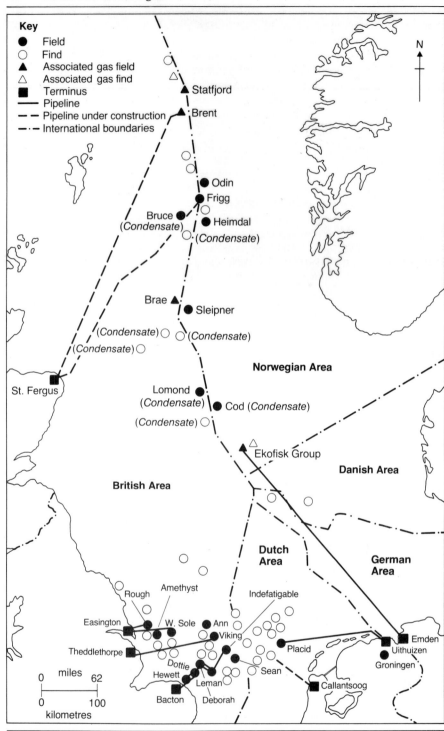

Fig. 3.2 North Sea gas discoveries

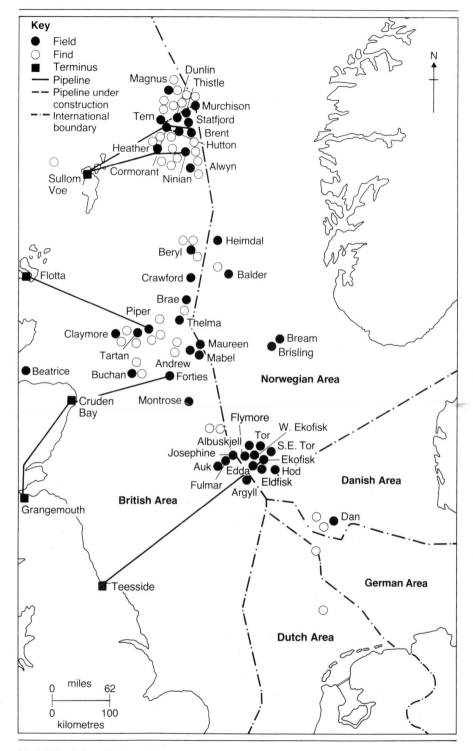

Fig 3.3 North Sea oil discoveries

Table 3.3 Number of exploration wells drilled 1968 to 1977 (*Source:* Department of Energy, *Development of the Oil and Gas Resources of the United Kingdom* 1978)

	1968	1969	1970	1971	1972	1973	1974	1975	1976	1977
East of England	30	34	12	7	8	7	4	2	3	5
East of Scotland	1	8	10	13	16	18	25	49	25	23
East of Shetland	—	—	—	4	8	16	26	23	25	24
West of England/ Wales	—	2	—	—	—	1	4	2	4	4
West of Shetland	—	—	—	—	1	—	8	3	1	11
Total	31	44	22	24	33	42	67	79	58	67

production levels stabilising from southern fields, increasing importance will be attached to discoveries in the northern sector. In the short term, supplies are guaranteed from the Frigg and Brent fields; indeed, the gas industry has amassed new industrial contracts to ensure that an over-supply problem does not occur.

The pricing of gas has been a controversial issue in recent years. Its competitiveness has caused the electricity industry to lose ground in industrial and domestic markets but suggestions that a tax on gas should be implemented have not been followed. The Gas Corporation maintains that the use of gas saves energy because it supplies markets which would otherwise use less efficient fuels. A factor which has created market distortions is the contract between ICI and British Gas in 1969 to supply 900 million therms a year for 15 years at 1.6p per therm.[22] Although both parties had reservations about the deal in 1969, the increase in energy prices in 1973/74 has given ICI cheap gas until 1984. British Gas attempted to invoke a hardship clause, but when the issue went to arbitration, ICI's case was upheld. Nevertheless, ICI may have to renegotiate a new contract as the Monopolies Commission expressed concern at the number of fertiliser producers forced into liquidation as a result of the cheaper availability of raw materials to ICI.

In the medium to long term, gas supplies to Britain are assured from Liverpool Bay which augment contracted supplies from the Frigg, Brent and Heimdal fields. However, Norway has still to find a market for its gas from the Statfjord and East and North East Frigg discoveries. The Norwegians have contemplated laying a pipeline across the deep trench to the Norwegian coast but the small domestic market may not justify the cost of such a project. It is more likely that the gas will be sold to Britain or other European buyers. In the past, the price of gas has been the main factor in dictating the destination of Norwegian production. Although Britain is geographically better situated for the landing of Norwegian gas by pipeline, the Ekofisk contract was placed with a continental consortium representing German, Dutch, Belgian and French consumers. The consortium was willing to pay a higher price than British Gas for the contract, which had an escalation clause relating future prices to those of other fuels. The Gas Corporation has had to accept higher prices with escalation clauses in its recent contracts: however, European consumers, in more urgent need of supplies, probably will pay a higher price to secure the Statfjord and other Norwegian contracts.

In the northern basins of the North Sea, many oilfields have a high proportion of gas in the reservoir. The oil companies require Government approval to flare this gas into the atmosphere to speed up oil production and the Government has recently taken a tougher line with the industry. In 1977 platform 'B' on the Brent field was non-operational because the Government insisted that gas re-injection equipment was installed on the platform to save gas. In some instances delays onshore have been responsible for flaring gas because gas separation plants have not been constructed to coincide with pipeline completion. At Sullum Voe, for example, the terminal was opened in November 1978, but the plant required for the separation of gases from crude oil will not be completed until March 1980. Because of this delay, an estimated £40 million to £80 million worth of gas will have to be flared from offshore platforms.[23] At St Fergus, methane is separated from hydrocarbon liquids – the heavier gases such as ethane, propane and butane, which in turn will be piped to Mossmorran in Fife, where a proposed gas separation plant will be constructed. Here, the propane and butane will be used as a fuel whilst ethane will be 'cracked' in a proposed ethylene manufacturing plant.

A possible alternative to flaring gas would be to liquefy the gas offshore, prior to its shipment to the UK or elsewhere by Liquefied Natural Gas (LNG) tanker. Ellerman City Lines and Ocean Phoenix Transport consider offshore liquefaction feasible and predict that a considerable amount of the 7,900 million cubic feet per day of associated gas produced in the 1980s will not be earmarked for removal by pipeline.[24] The sale of gas in liquefied form opens up possibilities for British Gas if the Government stops wasteful flaring of gas and encourages sufficient production to outstrip domestic demand. The World LNG trade of 21.5 bcm in 1977 (table 3.4) is expected to rise to 45 bcm in 1980 and 180 bcm by 1985.[25] (The US and Japan together will account for over 70 per cent of the demand).

Table 3.4 World LNG trade in 1977 (billion cubic metres) (*Source:* R. Dafter, *Financial Times,* 9 Nov. 1977)

Source	Quantity	Market	Quantity
Abu Dhabi	1.5		
Alaska	1.5	Japan	11.0
Indonesia	1.0		
Brunei	7.0		
Algeria	0.5	US	0.5
Algeria	6.0		
Libya	4.0	Europe	10.0
Total	21.5		21.5

Britain has experience of LNG as an importer of Algerian LNG, which is shipped to Canvey Island. The investment necessary for LNG is high because purpose-built tankers are expensive and transportation costs are seven times greater per ton than for the carrying of crude oil.[26] The 20 to 25 per cent higher calorific value per ton compared with crude oil partly compensates for increased costs and the British shipbuilding industry would benefit from the necessary carrier construction. In addition, the new employment created by the construction and operation of new terminals and liquefaction plants would not be inconsiderable.

At present Liquefied Petroleum Gas (LPG) is receiving more attention from the oil companies than LNG. Shell has won a contract to sell LPG valued at £535 million to the US over a ten-year period.[27] The 6 million tonnes of butane and propane involved will mainly come from the Brent field but whether the contract, which has price escalation factors built into it, is adhered to depends on the speed of construction of the Mossmorran plant. Originally, when the contract was signed, Shell had applied for planning permission to build the separation plant at Peterhead.

The level of involvement of the oil companies and British Gas in LNG and LPG production is problematic. If, for example, British Gas secures the Statfjord contract, supply could outstrip demand unless the Gas Corporation pays the oil companies to regulate production accordingly. In the medium term more gas can be acquired from marginal fields, especially in the southern sector, if higher prices are awarded. The prospect of future discoveries could further complicate long-term projections of demand and supply. In 1978 British Gas appears to be adopting a policy of dovetailing supply with demand to guarantee domestic customers supplies for as long as possible. Hence, there may never be surplus gas for the domestic market and the export of LNG may not be necessary.

British Gas also has reservations about the proposal to construct a gas-gathering pipeline system to minimise gas flaring in the North Sea. Engineering consultants Williams-Merz were commissioned by the Government in 1975 to examine the viability of a pipeline system. When the consultants reported in 1976, they proposed an intricate network covering most fields from Magnus in the north to Lomond in the south (Fig. 3.4). Over a twelve-year period and at a cost of £1.6 billion, Williams-Merz estimated that 1.5 billion cubic feet per day could be supplied to the grid and a further 6 to 9 million cubic feet of heavier gases could be delivered for the chemical industry or LPG production.[28] This report was designed to give an overview of the project and in the light of the consultants' comments the Government created a special company – Gas Gathering Pipelines Ltd (GGP) – to conduct a further investigation into the project's viability. The company's report, two years later, was less optimistic and recommended little modification to the existing pipeline network (Fig. 3.4). GGP felt that the network envisaged by Williams-Merz would be sub-economic as the pipelines would cost £5 billion to build and the landed gas price would be greater than that from the existing Brent and Frigg fields.[29]

A more logical but politically less practical proposal has been suggested by the consultants Buchanan and Clacher.[30] They favour an integrated pipeline network capable of carrying crude oil, natural gas and natural gas liquids from up to 60 fields from both the Norwegian and UK sectors. Such a system would minimise collection costs and make marginal fields more economically attractive. The consultants feel that both sectors could produce over 5 million barrels per day of oil, 12 billion cubic feet per day of natural gas and 30 million tonnes a year of gas liquids if the scheme was adopted. Although 12.5 per cent of hydrocarbon reserves in the North Sea contain gas liquids, most will be untapped because of the lack of a collection system.

In reservoirs with a high gas to oil ratio oil companies lose revenue by flaring gas. Buchanan and Clacher claim that if the ratio is 2,000 cubic feet of gas per

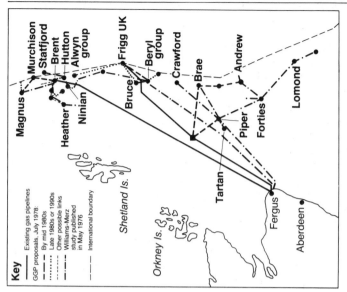

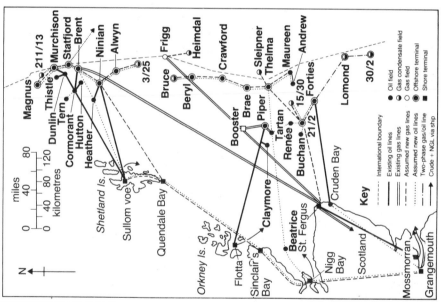

Fig. 3.4 Gas-gathering pipeline systems (*Source:* Buchanan and Clacher; Williams-Merz; and Department of Energy)

barrel, an operator wasting the gas content could lose about 30 per cent of potential revenue.

Pan Ocean, operating the Brae field, have found gas at a ratio of 3,000 to 7,000 cubic feet per barrel.[31] According to the consultant's criterion Pan Ocean should install equipment to handle gas and gas liquids. Surprisingly, Buchanan and Clacher costed their comprehensive network at £3 billion – £2 billion less than GGP's estimation for the Williams-Merz scheme. However, additional costs (£6 billion) would be incurred in the construction of new landing points (Nigg Bay and Quendale Bay) and a pipeline carrying natural gas liquids to the petro-chemical centres of Western Europe.

The indifference of British Gas to these proposals contrasts with the Government's willingness to carry out a gas gathering scheme. The Government is sensitive to the wasting of resources caused by flaring but is in a dilemma because if it insists on gas conservation, oil production could be impaired to the detriment of the balance of payments. Despite the GGP report, the Department of Energy has agreed with Shell-Esso on the practicality of a mini-system integrating the Cormorant, Heather, Ninian and NW Hutton fields with the main Brent pipeline. If further agreements are made on gas-gathering networks between the Department of Energy and offshore operators prior to the development of fields, the companies could plan the installation of appropriate systems to handle gas, oil or liquefied gas in advance.

The short-term prospect for British Gas is encouraging as the main problem facing the Corporation is the welcome one of matching supply with demand. The medium-term availability of supplies will depend on price – which could promote exploration and development of marginal areas – future oil discoveries in northern waters and new licensed areas, and the extent of a gas-gathering pipeline. In the long term natural gas production will decline and substitute natural gas (SNG), derived from coal, will be exploited. At Westfield Development Centre in Fife, British Gas, with Conoco collaboration, has gasified Pittsburgh coal, which is energy rich but difficult to use.[32] Commerical-sized plants are likely to be built in the US where SNG will be required sooner than in the UK.

North Sea oil

Within a decade of the early discoveries in 1969 and 1970, Britain will have become self-sufficient in oil. The intensification of exploration in the early 1970s in the central basin and Viking Graben reflects the early success of companies in locating major fields such as Brent, Forties and Ninian. On the UK continental shelf, oil operators have achieved a high ratio of success with one in every five exploration wells drilled producing a significant gas or oil discovery. The success ratio in the East Shetland Basin has been twice as great. The worldwide ratio is 1:20.[33] Exploration has declined since 1975 whilst the level of development drilling has built up over the same period (Tables 3.3 and 3.5). With the prospect of fewer major discoveries being made, the oil companies are concentrating their activities on the development of existing finds. The development programme is unfolding in two phases. The operators of the first generation of fields – the fourteen fields in Table 3.6 except Buchan, Tartan and Murchison – had announced development plans by the end of 1975. More than a year elapsed before Continental Oil announced its intention to develop the Murchison field. A new generation of fields is now in the process of development. In addition to Buchan and Tartan, development plans submitted to the Department of Energy for the Fulmar and Beatrice fields were due to be approved by the end of 1978 and BP, Shell/Esso and Pan Ocean may decide to develop Magnus, Cormorant North and Brae fields, respectively.

Table 3.5 Number of development wells drilled 1968 to 1977 (*Source:* Department of Energy, *Devel. of Oil and Gas Resources of the United Kingdom* 1978)

	1968	1969	1970	1971	1972	1973	1974	1975	1976	1977
East of England	36	27	28	34	36	21	20	13	7	7
East of Scotland	—	—	—	—	—	—	—	7	37	60
East of Shetland	—	—	—	—	—	—	—	1	10	29
Total	36	27	28	34	36	21	20	21	54	96

Table 3.6 Offshore oilfields in the North Sea (Spring 1978) (*Source:* Department of Energy, *Devel. of Oil and Gas Resources of the United Kingdom* 1978)

In production	Date of discovery	Date of commencement of production	Peak production year*	Peak production* (m. tonnes per year)†	Recoverable reserves* (m. tonnes)†
Argyll	Oct. 1971	June 1975	1977	1.1	—‡
Auk	Feb. 1971	Feb. 1976	1977	2.3	7.3
Beryl	Sept. 1972	June 1976	1978	4.0	75.0
Brent	July 1971	Nov. 1976	1982	23.0	220.0
Claymore	May 1974	Nov. 1977	1979	7.3	55.0
Forties	Nov. 1970	Nov. 1975	1978	24.0	240.0
Montrose	Sept. 1969	June 1976	1978	2.4	20.0
Piper	Jan. 1973	Dec. 1976	1979	14.6	82.0
Thistle	July 1973	Mar. 1978	1979	10.6	73.0
Under development					
Buchan	Aug. 1974	1979	—‡	—‡	—‡
Cormorant	Sept. 1972	1979	1981	3.0	15.0
Dunlin	July 1973	1979	1982	7.5	80.0
Heather	Dec. 1973	1978	1980	2.5	20.0
Murchison	Sept. 1975	1980	1982	7.2	51.0
Ninian	Jan. 1974	1978	1981	17.3	155.0
UK Statfjord	Apr. 1974	1979	1985	4.2	58.6
Tartan	Dec. 1974	1981	1981	—‡	—‡

* Operator's estimate

† 1 tonne = 7.4 barrels. 1 million barrels/day = 50 million tonnes/year.

‡ Under assessment.

As peak production levels in the largest fields begin to level off in the early 1980s (Table 3.6), the development of more small to medium-sized fields will become necessary in order to assure Britain of self-sufficiency throughout the 1980s. To achieve this level of output, further exploration activity in new areas is necessary. BP's oil discovery in the West Shetland Basin in 1977 has aroused interest in this area (Table 3.3) and perhaps the oil discovery onshore at Wytch Farm in Dorset will stimulate exploration in the newly licensed western approaches to the English Channel. The Dorset discovery has recoverable reserves of 50 million barrels, larger than the Argyll field and similar in size to Shell/Esso's Auk field. The relative cheapness of producing onshore oil may, therefore, encourage oil companies to investigate the possible landward extension of offshore reserves.

Costs, technology and marginality

The North Sea is a marginal area for the production of oil. The marginal costs of increasing production in Iran, Kuwait or Saudi Arabia would be 50 cents a barrel, bringing the delivered costs to a European refinery to $1.50 to $2.00 a barrel.[34] The operating costs in many UK fields are greater than this figure. Overall costs vary from $3 to $8 a barrel but since North Sea crude is light and low in sulphur content it can command a price of over $14 a barrel. Clearly, development costs will vary from field to field in the North Sea according to size, complexity of reservoir geology and distance from landfall. Generally, costs have risen for the most recent discoveries since exploration has moved to harsher marine environments where unpredictable weather and deeper water can lead to greater costs than originally forecast. Offshore operators, experienced in the shallow, calm waters of the Gulf of Mexico, used drilling ships and jack-up rigs to explore the southern North Sea. Unfortunately, in northern waters both types of technology proved unsuitable. Drilling days were lost which meant that money was lost, because neither type of exploration vessel could adapt to deep water or violent weather conditions. The charter of more expensive semi-submersible rigs increased costs by up to three or four times.[35] Similarly, cost estimates had to be revised upwards as production platforms, adapted from Gulf of Mexico technology, had to be re-designed to withstand the buffeting of North Sea wind and waves.

In the case of concrete production platforms, the central Ninian platform illustrates the engineering complexities involved in designing and constructing a structure which has to withstand waves of 100' and winds of 90 miles per hour. A 6,000 tonne steel deck had to be married to the 450,000 tonne, 500 feet high base prior to its 430 mile journey to the field.[36] The deck, built at Ardersier, had to be towed by barge around the Orkneys to the concrete platform yard at Kishorn. (See Fig. 3.5). Throughout its 72-hour trip, the wind had to be less than 20 knots and the waves less than 7 feet high; on its arrival, the concrete base was sunk with ballast and the deck was jacked to fit on top. During this operation, windless conditions were required, with waves less than a few feet high. The Meteorological Office in London earns £750,000 a year from weather forecasts supplied to North Sea operators.[37] The 'weather window' – a period of calms for towing platforms to fields, for supplying fields in production and for serving rigs and barges – is important to operators as lengthy periods of bad weather could delay the development programme.

Brown calculates the breakdown of development costs of a field with a pipeline as follows:[38]

Platform	30 per cent
Equipment for platform	10 per cent
Development drilling	15 per cent
Platform installation	15 per cent
Pipeline system	14 per cent
Land facilities and other costs	16 per cent

Fig. 3.5 Major British sites of onshore activity

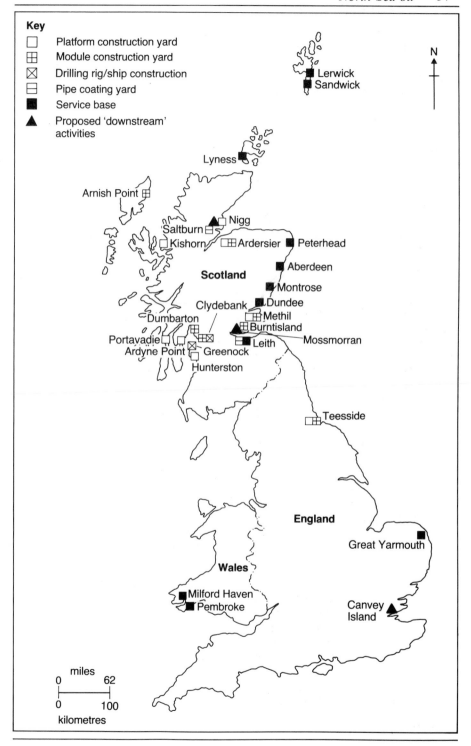

Key

☐	Platform construction yard
⊞	Module construction yard
⊠	Drilling rig/ship construction
⊟	Pipe coating yard
◼	Service base
▲	Proposed 'downstream' activities

Lerwick

Sandwick

N

Lyness

Arnish Point

Nigg

Saltburn

Kishorn

Ardersier   Peterhead

Scotland

Aberdeen

Montrose

Clydebank

Dundee

Dumbarton

Methil

Burntisland

Portavadie

Mossmorran

Ardyne Point

Leith

Greenock

Hunterston

Teesside

England

Great Yarmouth

Wales

Milford Haven

Pembroke

Canvey Island

miles

0   62

0   100

kilometres

Clearly a fixed platform, its installation and ancilliary equipment, accounts for a high proportion of total costs. As a result, operators who have discovered marginal fields seek alternative methods of recovering oil. The Department of Energy has disagreed with BP over the oil company's intention of using a converted semi-submersible rig to exploit its reserves (115–250 million barrels) from the Buchan field.[39] The Department fears that BP could 'cream' the field or abandon it if results were disappointing, whereas a commitment to the high costs of a fixed platform would ensure the thorough exploitation of reserves. Furthermore, the platform would have the capability for the installation of equipment for gas re-injection or natural gas liquid handling. Nevertheless, with many existing fields in the marginal category and the likelihood that future discoveries will be small to medium in size, operators will be using cheaper production methods. The BNOC, one of the partners in the Hutton field, intends to use a semi-submersible rig during the initial production stage before installing a more permanent tension leg production system.[40]

The North Sea environment has been largely responsible for the need to modify existing technologies and the consequent upward revision of costs. Even the Forties field, which had relatively few development problems, has incurred development costs of £800 million compared with original estimates of £300–350 million. To place this in perspective, Continental Oil's Murchison field has reserves of 51 million tonnes, about one-fifth of Forties, but development costs will be £475 million, and BP's Magnus field, with reserves of 60 million tonnes, has anticipated costs of around £1 billion.

The marginality of a field is not strictly determined by size although 500 million barrels is often deemed to be the cut-off point between profitability and marginality.[41] Two large fields – Ninian and Brae – are marginal high-cost discoveries because of their elongated shape and the complexities of their geological structure. Even though Ninian's reserves have been downgraded to 1.2 million barrels from original forecasts of 1.8 million barrels, development continued and the field came onstream in 1978. There may, however, be problems with the Brae field for which a development programme has yet to be submitted to the Department of Energy for approval. Appraisal wells show that flow rates vary to such a large extent that its reserves are difficult to estimate, but they are probably around a half of the original estimate of 500 million to 1 billion barrels.

An accurate estimate of the size of reservoir is imperative if an operator is to assess the viability of constructing a pipeline to the shore. The cost is justified in most of the major discoveries which are linked by pipeline to locationally convenient transhipment points (Sullom Voe and Flotta) or the nearest landfall site (Cruden Bay). However, most small fields – Montrose, Auk, and Argyll – use an offshore loading system. Mesa Petroleum wished to incorporate a similar system in the development of its Beatrice field but the Department of Energy will only give approval if a pipeline is used. Mesa had claimed that the 'waxy' nature of the oil would make it unsuitable for pipeline transportation. However, the size of the field – around 450 million barrels – could economically justify the construction of a pipeline and minimise the danger of a major oilspill during tanker loading at a site only twelve miles from the coast.

Government involvement

Government involvement in the development of North Sea resources has increased in the 1970s as offshore activity moved from the exploration to the production stage. Initially, the Government's main concern was with the allocation of licences but, as the oil began to flow ashore, a new system of taxation and control was devised to maximise the benefit to the British economy.

The Government offered blocks in six licensing rounds in 1964, 1965, 1970, 1972, 1976 and 1978. Production licences for blocks were awarded on a discretionary basis, except for fifteen blocks in the fourth round, when an auction system was used, It can be argued that greater revenues are derived through the latter method, but the discretionary system does seem to favour domestic companies; for example, it is perhaps not coincidental that BP has been awarded prime blocks. The Government has allocated licences on the submission of work schedules with the provision that at least half the area must be returned within six years. The increase in the number of exploration wells in 1977 can be attributed to the acceleration of the oil industry's exploration programme before the compulsory return of unwanted blocks allocated in 1971.

In the early licensing rounds most of the British North Sea was offered to the oil companies for licence applications. This was understandable because little was then known of the size of the area's reserves. Chapman has claimed that after the unexpected successes, especially in northern waters, the Government could have taken a more cautious view in the fourth licensing round.[42] It is now Government policy to offer a small number of blocks at more frequent intervals. Nevertheless, the generous approach of the past has now left relatively few areas which appear attractive to the oil industry. Unlike Norway, Britain has run out of 'golden blocks' to allocate and exploration activity has slumped considerably in 1978, with only 16 rigs in operation compared with 30 in 1977.[43]

The prospect of unexpected revenues from the increasing number of discoveries forced the Government to re-assess its position with respect to the North Sea. In 1974, the Department of Energy announced its intentions in a White Paper; it hoped to assert greater public control and give a fair share of profits to the nation.[44] These proposals were incorporated into the Oil Taxation Act 1975 and the Petroleum and Submarine Pipe-lines Act 1975. In 1973 the House of Commons Public Accounts Committee had pointed out that British companies, under existing tax legislation, could offset their growing liabilities to foreign oil-producing governments against North Sea profits.[45] Also, the rate of return, after tax, on capital invested in the North Sea would be as high as 60 per cent in some cases, compared with $8\frac{1}{2}$–9 per cent for British manufacturing industry as a whole.[46]

The Government had to introduce new fiscal legislation which would continue to give the oil companies a fair return on investment without letting them make excessive profits. The three taxes introduced were:

(a) A royalty of $12\frac{1}{2}$ per cent on the well head value – this could be taken in kind;
(b) Petroleum Revenue Tax (PRT) of 45 per cent;
(c) Corporation Tax of 52 per cent.

The new taxation system only applies to the UK continental shelf, and PRT to each individual field. Companies are subject to PRT after deducting royalty payments, operating expenditure, $1\frac{3}{4}$ times capital costs and 100 per cent of abortive exploration expenditure. A total of 1 million tonnes per field per year is also exempt up to a cumulative maximum of 10 million tonnes.

Cochrane and Francis calculate that an oil company operating an average field would lose 77 per cent of its profits to the Government.[47] Nevertheless only the low-cost larger fields such as Forties and Piper will be subject to PRT; companies operating in the smaller fields or high-cost fields will be able to write off capital expenditure or the oil allowance per annum against profits. Even the operators of larger fields will pay little tax in the early years as they recoup their development costs. In 1977 the gross trading profits from the North Sea amounted to £1.7 billion – an eighth of all industrial and commercial profits in the UK. The tax paid was £50 million.[48] It is with this background that the Labour Government in 1978 was planning a substantial increase in PRT, possibly to 60 per cent, to compensate for the delays in the paying of tax.

The main concern expressed by the oil companies is not over tax but state participation in oil developments. Although most companies operating in the North Sea are UK based, the majority of them are foreign-owned. The discretionary system of licensing has increased British company involvement from 30 to 43 per cent between rounds 3 and 4.[49] Naturally, participation by UK companies varies from field to field; for example, BP has 97 per cent share of Forties but no UK company is involved in the development of the Heather field.[50] The Labour Government tried to persuade companies to enter joint agreements with the NCB and British Gas (then, the Gas Council), in the early licensing rounds. This policy was shelved during the Conservative Administration from 1970 to 1974. On the return of a Labour Government in 1974, a commitment to greater public control was realised with the creation of the BNOC in 1976 under the terms of Part 1 of the Petroleum and Submarine Pipe-lines Act 1975.

The BNOC both acts as an oil company in its own right, and also advises the Secretary of State on oil matters as a contribution to national policy decisions. The company has taken over the interests of the NCB and in 1977 concluded participation agreements with 27 companies and outline terms with 15 which had discovered commercial fields under existing licences.[51] This gives the BNOC access to over a half of UK oil production by 1981 and allows Britain to play a greater role in the destination of oil products, that is, moving towards the role of British Gas as a sole buyer.

The Government has used the participation issue as a lever in the fifth licensing round, excluding Amoco because of its reluctance to come to an agreement with the BNOC. In the fifth round, the BNOC was awarded a 51 per cent interest in seven production licences and, as the company builds up its staff expertise, it will become a major operator in the 1980s.[52] The expansion of the BNOC and its role in exploration and development may well be necessary in the 1980s, as other oil companies may move to non-British waters because of the issuing of the best blocks too early in the North Sea development programme. It seems probable that the BNOC will take the lead in exploring new licensed areas.

The Government's depletion policy, also incorporated into the 1975 Act, could discourage exploration in the near future. The policy is flexible and discoveries made before 1975 under the fourth and earlier licensing rounds, would not be subject to controls until 1982 or four years after production commenced, whichever was the sooner.[53] Even post-1975 discoveries, if they were made under pre-1972 licensing arrangements would not have to cut production until one and a half times the capital investment was recovered. The Department of Energy is now adopting a stricter attitude to the wasteful flaring of gas and is thoroughly scrutinising new development plans. Both measures could result in a slower build-up of oil production, and after 1981 the Government could ask operators to reduce output by as much as 20 per cent. The extent to which this legislation is enforced will depend on how the competing aims of self-sufficiency, improving the Balance of Payments and prolonging the life of North Sea oil reserves are reconciled.

The offshore market

The development of North Sea oil and gas resources has opened up new opportunities for British industry to supply goods and services offshore. Industry has been slow to respond to new market demands but the British share in the offshore market has increased with the help of the Offshore Supplies Office (OSO). This organisation was created after a report in 1973 by the International Management and Engineering Group which estimated that a £300 million-a-year market could be exploited in the UK sector of the North Sea, but that only 25 to 30 per cent of this market would be captured by British companies on then current trends.[54] OSO was therefore created to act as an agent between operators and suppliers. This role was reinforced in 1975 when the Government adopted a 'buy British' policy and issued a Code of Practice for OSO and the Offshore Operators Association, obliging operators to use British goods and services when tenders were competitive in terms of price, delivery, service and specification.

As a result of Government intervention and the awakening of industrial companies to North Sea opportunities, the UK share of a stabilising market has gradually risen from 40 to 62 per cent in 3 years (Table 3.7). The character of the offshore market is changing as oil companies move from the exploration to the development and production stage. This is reflected in the value of orders placed for production platforms, modules and rig hire for exploration and development. Rig hire made up 11½ per cent of all orders in 1974, 5½ per cent in 1977, and the fabrication industry (platforms and modules) accounted for 34 per cent of orders in 1974, 23 per cent in 1977. Whereas foreign companies dominated the rig hire market, the Government has encouraged the new UK fabrication industry to secure a greater share of the North Sea market. Unfortunately, the size of this market seems to have been grossly over-estimated. In 1974 the Department of Energy estimated that up to 80 production platforms would be required by 1980.[55] In July 1978 only 24 platforms had been installed and a forecast of 30 platforms would appear more realistic for 1980. The deceleration of development progress in the mid-1970s and a move to cheaper, alternative production techniques are having repercussions throughout the fabrication industry, especially in the manufacture of concrete platforms.

The early orders for concrete platforms were lost to Norwegian yards because of planning delays in the choice of site. After the rejection of a planning application for a yard at Drumbuie, the Government overreacted and in an attempt to build up production capacity subsidised the construction of two yards at Hunterston and Portavadie (Fig. 3.5). In 1978 fabricators specialising in concrete platforms were without new orders including the Government-sponsored yards which have never yet won an order.

Table 3.7 An analysis of orders placed 1974 to 1977 (*Source:* OSO Annual Reports)

	Value of orders placed (£ million)	UK share	%
1974	1279	516	40
1975	1185	613	52
1976	1041	591	57
1977	1295	806	62

The prospect of continued orders for steel production platforms is not encouraging. Laing Offshore at Teesside had not won an order for over two years (by the end of 1978) and Redpath Dorman Long (RDL) ran its yard at Methil on a care and maintenance basis for over a year until it received a share of the Texaco contract for the Tartan field in September 1977. It is believed that UIE of Le Havre offered a tender £5 million below the RDL bid but Government pressure was placed on Texaco to split the order to alleviate unemployment in the Methil area.[56] Subsequently, RDL amalgamated with de Groot of the Netherlands to form Redpath de Groot Caledonian and has secured two small orders which will guarantee employment until late 1979.[57] Highlands Fabricators at Nigg Bay and McDermott at Ardersier have been more successful in maintaining continuity of orders. Highlands Fabricators have specialised in the construction of giant platforms for two of the largest fields at Forties and Ninian. McDermott has won small export orders for the Dutch and Brazilian offshore areas. Orders such as these might be the only possible salvation for some of these yards because the home market will generate only a trickle of major orders – perhaps between five and ten – from 1979 to 1985.

Although the performance of British industry has improved in recent years, the shipbuilding sector has not diversified for the North Sea market to the same extent as have our European competitors. Norway, France, the Netherlands and Germany have captured a major share of the offshore market for drilling rigs and ships, pipe-laying barges and supply vessels which compensates for their loss of the tanker market since the oil crisis.[58] Unfortunately, in Britain only the Clydeside shipbuilders have broken into the offshore market and their contribution is small compared with the output of rigs and vessels from continental yards (see Fig. 3.6). The Government, to prevent unemployment, is financing the construction of a £14 million drilling rig for the BNOC at the Marathon yard, Clydebank. This rig, like the others that have been built on Clydeside, is a jack-up rig which is not suited for the northern North Sea. The BNOC will have to charter the rig in shallower offshore waters. In contrast, European competitors, especially the Aker Group in Norway, have been involved in the design and construction of semi-submersible rigs for harsher marine conditions. In the provision of supply boats, the Swedes, Norwegians and Germans have established a firm foothold in the market and shipbuilding tonnages reached new records, whilst British yards failed to maintain employment during the recession.[59]

As the North Sea oil fields develop, the emphasis in the offshore market will move from the supplying of platforms, rigs and other capital equipment to the provision of services. One aspect which will increase in importance throughout the 1980s is that of offshore maintenance. In the Gulf of Mexico, regular inspection and maintenance was of course necessary and in some instances

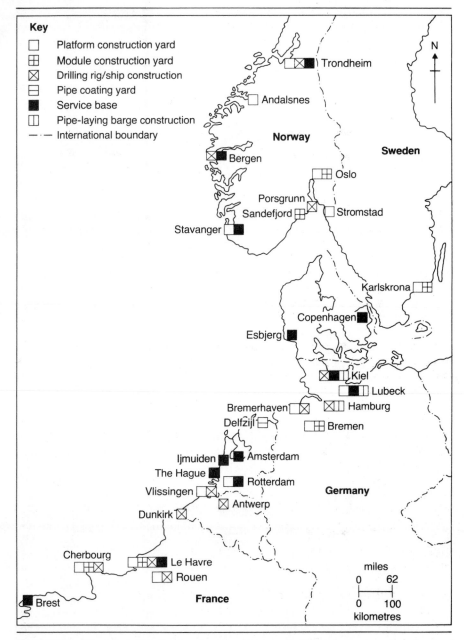

Fig. 3.6 Major European sites of onshore activity. (*Source:* after E. de Keyser, *European Offshore Oil and Gas Yearbook 1975/76*)

platforms were abandoned or replaced. Even so, the cost was minimal – around 2 cents a barrel; on the other hand, inspection and maintenance of North Sea

structures could cost 60 to 100 cents a barrel and provide orders valued at
£300–£400 million a year in the 1980s for firms offering this service.[60] All
structures in the North Sea must be inspected and certified for safety under the
Offshore Installations Regulations. From the limited amount of contracts to
date, British companies have secured around two-thirds of the market. The
creation of OMISCO (Offshore Maintenance and Inspection Company) com-
bines the offshore experience of BP and George Wimpey, but other British
consortia could be formed to take advantage of this sector of the offshore
market in the 1980s. With US groups having little experience in this field, an
opportunity exists for British industry to develop expertise in the North Sea
which can then be exported to other offshore areas.

Regional impacts

By 1976 100,000 jobs had been created directly and indirectly in the UK for the
provision of goods and services for the North Sea market.[61] The greatest
impact has occurred in Scotland. Fig. 3.5 shows the major areas of onshore
activity. The Grampian, Highland and Islands regions offer the best geographical
locations for service bases, platform construction sites and landfall terminals
for offshore pipelines. Nevertheless, 30 per cent of all *direct* employment is in
the Strathclyde region: 6,750 jobs have been created there by firms which have
devoted part of their business to supplying equipment for offshore operators
(Table 3.8). New employment in the Highland and Grampian regions was
mostly created by newly established firms wholly committed to serving the
offshore market. In the Highland region most of the oil-related workforce is
involved in the construction of platforms and modules at Nigg and Kishorn. On
the other hand, most of the employment generated in the Grampian region has
been in the service sector. The 'oil capital', Aberdeen, and its satellites at
Peterhead and Montrose (Tayside region) provide employment in marine
transport, catering, warehousing and administrative back-up services.

Unfortunately, the distribution of oil-related jobs has not favoured the areas
in greatest need of employment. The Grampian, Orkney and Shetland regions
had low unemployment rates, well below the Scottish average, prior to the
build-up of oil-related employment (Table 3.9). By contrast, unemployment in
Strathclyde has consistently been above the Scottish average. From 1970 to
1976 around 60,000 jobs (including indirect employment) have been created in
Scotland but from 1966 to 1972 the same number of jobs were lost through the
contraction of traditional industries – coal mining, textiles, shipbuilding and
steel-making. Most of the 'lost' jobs were in the industrial central belt, especially
Clydeside. Oil-related employment has been valuable to Strathclyde, but it
only represents 1 per cent of the total employment in the region compared with
7 per cent and 9 per cent in the Grampian and Highland regions respectively.
The 'oil boom' has barely transformed the Scottish economy. In terms of the
classical economic indicators of out-migration and unemployment, Scotland is
still a depressed region. Out-migration has continued throughout the 1970s; it
fell from 27,600 in 1972 to 2,000 in 1974, only to rise again in 1975 to 19,000.
The unemployment situation improved from 1973 to 1975 but it has subsequently

---

Table 3.8 Oil-related employment in Scotland (*Source:* Scottish Office: *Scottish Economic
Bulletin*, No. 11, Winter 1977)

Region		Oil employment in: Wholly* involved units (1)	Partly† involved units (2)	All units (3)	Change since 1974 Employment (4)	% (5)	Survey oil jobs as percentage of total employment in region (%) (6)
Strathclyde	Manufacturing	4,200	6,750	10,950			
	Non-manufacturing	100	400	500			
	Total survey	4,250	7,150	11,400	+ 150	+ 1	1.1
Fife	Manufacturing	2,000	250	2,250			
	Non-manufacturing	—	—	50			
	Total survey	2,000	250	2,250	+ 600	+37	1.9
Tayside	Manufacturing	400	150	500			
	Non-manufacturing	800	100	850			
	Total survey	1,150	200	1,350	+ 400	+39	0.9
Grampian	Manufacturing	1,050	1,850	2,900			
	Non-manufacturing	9,300	450	9,750			
	Total survey	10,350	2,300	12,650	+4,000	+47	7.3
Highland, Western Isles, Orkney and Shetland	Manufacturing	7,300	50	7,300			
	Non-manufacturing	600	50	650			
	Total survey	7,850	100	7,950	+3,900	+96	9.3
Lothian, Central, Borders and Dumfries and Galloway	Manufacturing	550	1,200	1,750			
	Non-manufacturing	450	150	600			
	Total survey	1,000	1,350	2,350	− 650	−22	0.5
Total Scotland	Manufacturing	15,450	10,200	25,650			
	Non-manufacturing	11,200	1,150	12,400			
	Total survey	26,650	11,350	38,000	+8,400	+28	1.8

(Due to rounding, the totals may differ from the sum of the individual components)

* Firms producing solely for the North Sea market
† Firms producing partly for the North Sea Market

deteriorated (Table 3.9). In relation to the UK unemployment rates however, Scotland's position has improved in that this rate was double the UK average in 1965 and only 25 per cent higher in 1977.

Table 3.9 Unemployment rates in the regions of Scotland (*Source:* Scottish Office: *Scottish Economic Bulletin,* Nos. 11 and 15)

	1970	1972	1974	1976	1978
Scotland	3.8	6.0	3.6	6.7	7.7
Highland	4.4	4.8	3.0	5.9	7.7
Grampian	3.0	3.6	1.8	3.3	4.1
Tayside	3.7	5.7	3.0	6.3	7.0
Fife	4.1	5.6	3.3	6.4	7.1
Lothian	3.3	4.4	3.1	5.5	6.3
Central	2.8	5.3	3.2	6.1	6.5
Borders	1.8	2.3	1.4	3.7	3.8
Strathclyde	4.5	7.2	4.5	7.9	9.3
Dumfries and Galloway	4.9	5.2	3.7	7.4	7.6
Orkney	4.2	2.6	2.3	3.4	4.4
Shetland	4.1	3.7	2.2	3.5	2.9
Western Isles	15.3	17.8	11.9	14.5	13.3

MacKay and MacKay predicted that *direct* oil-related employment would peak in 1976 at 27,000 declining to 21,400 in 1980.[62] Once the production stage is reached, oil-related employment tends to decline since fewer personnel are required in the maintenance and operating of platforms, rigs and terminals than were necessary in their construction. The hiatus in platform construction has resulted in a slower build-up of employment in that sector than envisaged by MacKay and MacKay. Nevertheless, the number of direct oil-related jobs has been greater in overall terms than their forecast, increasing from 18,250 in 1974 to 26,650 in 1976 and 30,300 in 1978. It is likely that the employment figures will begin to fall in 1979 and 1980 as exploration activity declines and unemployment in platform yards, which employed substantial workforces, increases. The April 1978 figures highlight the problem of regional imbalance in oil-related employment (Table 3.10). In two years the Grampian region has consolidated its dominance at the expense of Strathclyde where only 500 wholly oil-related jobs remain as a result of the run-down of employment at the Ardyne platform yard. This pattern will be reinforced as the oil industry moves into the production phase. The market for capital equipment is diminishing and the east coast regions are best placed to capture the offshore maintenance market because of their proximity to the oilfields.

Table 3.10 Wholly oil-related employment in Scotland, April 1978 (*Source:* Scottish Office: *Scottish Economic Bulletin,* No. 15, Summer 1978)

Strathclyde	500
Fife	1,400
Highland	7,600
Grampian	17,400
Tayside	2,000
Island Regions	800
Others	600
Total	30,300

Refining and petrochemicals

The Government hopes to create further employment in the UK by means of 'downstream' activities, that is, the refining and use of oil in the petrochemical industry. North Sea oil is light, low in sulphur content and therefore commands a high price because it can be used in the manufacture of high-value products such as petrol and naphtha (for the chemical industry). All oil and gas produced in the UK sector of the North Sea must be landed in Britain and it is the Government's intention that two-thirds of the oil shall be refined in the UK.[63] This policy has been questioned by the oil companies, which argue that only one-third of the crude oil refined in Britain need be of North Sea quality. Consequently, they claim, the surplus of this high-priced crude should be exported whilst Britain continues to run its refineries on cheaper, imported heavy oil from the Middle East. On the other hand, the Government would prefer the country to be self-sufficient from domestic resources and to export these high-value refined products rather than crude oil.

It is against this background that the Government approved Cromarty Petroleum's proposed refinery to handle North Sea oil at Nigg. Occidental are also planning a new refinery at Canvey Island for the same purpose. These proposals have caused some unease in the oil industry because of the existing refining overcapacity in Britain. In the EEC as a whole, refineries are operating at only 63 per cent of capacity and it is possible that the EEC Commission will recommend that community aid and regional development grants from member governments should not be applicable to refinery projects until the over-capacity situation improves.[64] This could have repercussions on the Nigg refinery, where 40 per cent of the total cost of the project will be met by various Government grants.

The further development of the petro-chemical industry is another 'downstream' activity which could benefit from the availability of North Sea oil and gas. The combined capacity of UK petro-chemical plants, producing primary materials, is around 8 million tons a year or 12.8 per cent of total West European capacity. There is ample opportunity for the UK chemical industry to secure a greater share of the plastics, organic chemical and ethylene market.[65] The Scottish Development Department has earmarked six areas for potential petro-chemical industrial development and has advised local authorities to investigate suitable sites in these areas, giving due consideration to environmental factors.[66] Only one out of the six chosen areas is not in the central belt – the Cromarty Firth, where a refinery is due to be constructed. The other areas – north Ayrshire, east Glasgow, Bathgate, Grangemouth and Central Fife – are all served by good communications. Already one of the nominated areas – Central Fife – is likely to be approved by the Secretary of State for the construction of a LPG tanker jetty at Braefoot Bay and a gas separation plant and an ethylene cracker at Mossmorran. Central Fife, like the other central belt areas, suffers from high unemployment and perhaps the petro-chemical industry can help to redress the regional imbalance of oil-related employment from the Grampian and Highland regions to the central belt regions.

Planning for oil

The rapid impact of North Sea oil activity took planning authorities by surprise.

Areas which had been planning for slow industrial growth and population decline were suddenly faced with planning applications for platform construction yards and oil and gas terminals. Remote sites often offered the best location for fabrication work; for example, concrete platform yards need to be located in sheltered, deep-water zones. The rejection of a planning application for such a yard at Drumbuie – the best site in Scotland to fit these locational requirements – emphasises the conflict between the planning needs of the oil industry and those of small communities. The Highland and Grampian regions have the best geographical locations for oil-related activity; Central Scotland, on the other hand, is not a convenient location for the oil industry but it has a skilled labour force and a social and economic infrastructure that is lacking in the northern areas. After the Drumbuie inquiry, the Scottish Development Department produced its coastal planning guidelines. These indicated preferred conservation areas and preferred development zones in an attempt to channel development to areas with sizeable existing populations.[67] The guidelines were somewhat belated as planning permission had already been given for many oil-related developments, including the Howard-Doris application to construct concrete platforms at Loch Kishorn, ten miles north of Drumbuie.

The greatest planning problem related to large-scale developments in rural areas such as Nigg and Kishorn is that much of the workforce are temporary immigrants drafted in on high wages for a short term to complete a contract on time. It is therefore extremely difficult for the planning authority to estimate the extent of infrastructure services required; new houses, schools and shops may become redundant with the completion of the work.

The north-east of Scotland has been most affected by onshore developments. Aberdeen, the 'Dallas of Europe', was inundated with planning applications for quay space, warehousing and for development in all sectors of the property market. Unlike towns dependent on construction activity, Aberdeen is assured of continued employment in oil but the influx of the oil companies has created imbalance in the local economy. This was aggravated by the Government's incomes policy from 1972 to 1974 when local firms were obliged to follow Pay Board guidelines whilst the oil companies as new employers were not bound by incomes policy. Labour shortages occurred in both the old-established private and public sectors as oil firms attracted their employees with higher wages. Property prices rose as pressure on housing, offices and retail space intensified. Criticism has been levelled at big business in Aberdeen because the oil boom seems to have benefited the wrong elements of society.[68] Taylor has questioned the pace of development in Aberdeen and Peterhead and in particular has lamented the destruction of the old fishing settlement, Old Torry, in Aberdeen.[69] Similarly, the rush to develop the harbour area in Peterhead has resulted in a tourist attraction formerly nicknamed affectionately the Harbour of Refuge being ironically re-christened the Harbour of Refuse.[70]

It can be argued that the oil industry had few constraints placed on it by planning authorities on the mainland, but this is far from the case in Orkney and Shetland. In Orkney, the Island Council negotiated with Occidental over its proposal to construct a landfall terminal at Flotta. Clearly, the Orcadians were concerned about the possible environmental disruption which could occur during the construction and operation of a terminal planned to handle 400,000 barrels of oil per day. As a result, the operators have agreed to pay 2p per ton 'disturbance allowance' as a royalty for every ton shipped through the terminal in addition to harbour authority payments.[71]

In Shetland, the County Council has tried to gain the greatest benefits through hard bargaining with the oil companies. The island, rich in history, flora and fauna, had benefited from Highlands and Islands Development Board investment in the 1960s to revitalise its ailing economy. By 1971 Shetland was beginning to prosper with half of the island's labour force employed in traditional industries – agriculture, fishing, fish processing and knitwear.[72] The discovery of oil east of Shetland meant the development of Lerwick as a service centre for the oil industry and Sumburgh as a helicopter base and major airport.[73] The main zone of activity, however, is at Sullom Voe, which will be developed as an oil terminal. In the Shetland County Council Bill 1974, the Council received wide powers to control developments around the Voe. The Council argued strongly that a Reserve Fund should be created to ensure future prosperity for Shetlanders after the oil boom had subsided, and the oil companies have agreed to pay £25 million to the local authority by 2000.[74] In addition to the Reserve Fund, port and harbour authority payments should amount to 1p per ton of oil passing through the terminal. At times, the partnership between the Council and oil companies has been uneasy and the Sullom Voe Association, through which both parties have an equal share in the terminal's development, has often been the focus of disagreement. Indeed, the delay in the construction of the terminal can partly be attributed to the haggling between the authority and the oil industry over the design and location of the storage tanks.

Ultimately, Shetlanders will benefit from oil. Traditional industries are not economically linked to oil development and should survive even after the terminal has become obsolete. Nevertheless, despite the attempts of the Council to 'quarantine' oil activities, social and economic problems have inevitably arisen. Local industries and services have suffered labour shortages because they cannot compete with the exceptionally high wages paid at Sullom Voe.[75] The knitwear and fishing industries experienced a relative decline in 1975 and lost employees to the oil industry. They then could not attract them back on lower salaries when business began to improve. The problems of affluence are reflected in an increased incidence of crime, drunkenness and road accidents. This has also occurred on the mainland, but many Scottish authorities may ask themselves why they have had to contend with similar socio-economic problems without the benefits of guaranteed revenues for the future through a 'disturbance allowance' scheme.

Environmental implications of North Sea oil and gas development

The exploration and development of North Sea oil and gas resources has inevitably increased the level of activity in a marine area that has always been a main thoroughfare for shipping. Kenward, on visiting the Brent field, described it as a hive of activity. Congestion in the sea-lanes is aggravated by up to 50 vessels per month serving the rigs and the number of helicopter movements per month compares with the intensity of air traffic around Heathrow.[76] The marine environment of the North Sea is one of the most difficult yet encountered by the oil industry. Fatalities per thousand employees are higher than in any other fuel industry, and the North Sea had claimed the lives of 88 offshore employees up to 1977.[77]

The North Sea is a breeding ground for large numbers of birds and fish. These are threatened by an activity which can only add to the oil pollution problems around our shores. This is recognised by the Nature Conservancy which commissioned the Institute of Terrestrial Ecology of the Natural Environment Research Council to survey the ecology of the Shetland Isles in order to plan for the preservation of the archipelago's rare flora and fauna.[78]

Are proper precautions being taken to minimise environmental disruption? The Government has made it safer to work offshore through the Health and Safety at Work Act and a 'certificate of fitness' is required from the Department of Energy before the oil companies can operate offshore installations. Nevertheless, fears have been expressed that some platforms may not last the lifetime of their fields. No structures elsewhere have ever had to withstand the depth of water or the velocity of wind and wave encountered in the northern waters. Even the scouring and corrosion of platforms in the southern waters has been much greater than anticipated.[79] The structures in the north are regularly inspected and maintained, but the scale of the problem may be insurmountable. The supervising engineer of Comex has commented that the task of inspecting and maintaining a large concrete platform can be compared with finding a crack in a Heathrow Airport runway one night with a torch and then going back the next night to repair it.[80]

The development of platforms, rigs and pipelines in the North Sea has produced inevitable conflict with the fishing industry. Fixed installations are protected by a 500-metre zone from vessel infiltration, but 260 incidents were reported to the Government from 1968 to 1976 involving ships moving too close to installations and in some cases causing damage.[81] Trawlers are the main culprits; however, the fishermen are resentful of the oil industry. Fish tend to cluster around platforms because of the availability of food but trawlers are legally required not to follow them. The oil industry has also been responsible for depositing debris on the seabed which has torn or destroyed nets. The laying of the Brent and Ninian pipelines exemplifies this point; by May 1976, 24 claims for compensation had been made by fishermen of which 16 were upheld.[82]

The main environmental problem which concerns the fishermen is not that of oil-related debris but the threat of additional oil pollution from North Sea activities. The threat of blow-outs, pipeline failure, structural faults and oil spills during the loading of tankers and the transporting of oil could have widespread repercussions for the ecological balance of the North Sea. Red Adair, the blow-out expert, claimed in April 1977 that the safety precautions against a major blow-out in the UK were negligible.[83] Two weeks later, the Ekofisk platform, Bravo, experienced a blow-out when pressure built up through Well 14 and the blow-out preventer did not operate. Up to 4,000 tonnes of oil per day poured into the sea until Adair's team capped the well on the eighth day. In spite of the Department of Environment's prediction that there is a 50 per cent probability of a blow-out by 1981,[84] the offshore operators have been complacent about this issue. Shell computed 5,000 likely blow-out accidents and estimated that oil would only reach the coast in 10 per cent of all cases.[85] The company calculated that if a blow-out occurred in the Ninian field no coastal pollution would result because of the nature and direction of North Sea currents and prevailing winds.

This type of analysis by Shell is questionable since it implies that pollution at sea is acceptable if it does not reach the coast. It also discounts the possibility of a platform catching fire. In Bay Marchand in the United States, Adair took months to control a fire on a small platform with only a fraction of the number of wells of many of the larger structures in the North Sea. Since the Ekofisk blow-out, improvements in fire-fighting equipment have been made with the introduction of the ubiquitous 'Uncle John'. This semi-submersible can be chartered for pipe-laying or fire-fighting. BP are building a semi-submersible which will be more efficient in handling blow-outs than its Forties Kiwi which aided Phillips during the Ekofisk incident. Adair would undoubtedly class these measures as barely adequate, and he is perhaps justified in his assertion that safety is a low priority in the allocation of investment for oil development.[86]

The transportation of oil and gas constitutes a potential environmental hazard. Pipeline failures seem inevitable because of incomplete knowledge of the seabed, and doubts have been expressed whether all of the pipes have been laid squarely in the trenches prepared for them.[87] If a pipe was laid skew across the edge of a trench, a greater chance of a failure would exist because of stress exerted on the pipe. If a pipeline break occurred, the automatic shut-off system would not be able to detect losses up to 2 per cent of the daily output. Johnston estimates that it is possible for 30,000 barrels of oil per day to leak undetected from the Ninian and Brent pipeline systems.[88] To minimise the risk of any leak being undetected, greater surveillance of pipeline routes could be undertaken. Aircraft could also patrol these areas to ensure that there are no ships anchored or fishing trawlers present over the pipelines which might damage them.

The movement of oil and gas by tanker will increase as peak production is reached, adding to existing shipping congestion in the North Sea and English Channel. The storage and transportation of LNG and LPG offers the greatest risk of accidents. If a LNG tanker ran aground or was involved in a collision, it has been estimated that the cargo if released would burn at the rate of 10,000 tons per minute with a high risk of explosion.[89] At Canvey Island, the British Gas LNG terminal presents the highest risk in this complex of oil and gas 'downstream' installations and the Health and Safety Executive has recommended the resiting of the terminal at least four kilometres away from housing areas.[90] In the Netherlands and California, LNG storage and handling facilities must be located 32 and 6½ kilometres respectively from populated areas.

The question of the proximity of the Mossmorran project to the towns of Dalgety Bay and Aberdour has provoked most opposition to Shell-Esso's proposals there. The proposed LPG terminal at Braefoot Bay is only two kilometres away from housing areas in both towns. The terminal will be connected by a six kilometre pipeline inland to Mossmorran where the gas separation plant will be constructed. The problem in Mossmorran is compounded and the risk intensified because of the number of potentially hazardous plants proposed within the same area. In addition to the marine terminal and separation plant, Esso Chemicals intend to build an ethylene plant at Mossmorran.[91] The fears of local residents are understandable in view of the accident which destroyed a natural gas separation plant in Qatar, killing seven people, in April 1978. The applicant, Shell, was involved in the design, construction and operation of this plant and in the light of the accident is modifying the design of the proposed Mossmorran installation. The Secretary of State has delayed his final approval because of the risks presented by a radio transmitter at Braefoot Bay. It is likely that measures will be taken to re-site the transmitter, which could relay current to an aerial – the petro-chemical complex – and cause an explosion.

A more immediate problem is that of oil tanker accidents and their consequences. The English Channel is the busiest sea lane in the world and one third of the world's shipping collisions occur here. Oil and gas activity in coastal waters – drilling in the western approaches of the Channel and increased transportation of oil – can only increase the risk of future accidents. During the 1970s tonnage lost has increased each year; 1977 was the worst on record and the outlook for 1978 was expected to be no better.[92] In March 1978 the *Amoco Cadiz*, carrying 230,000 tonnes of crude oil, ran aground near Portsall and polluted long streches of the north and north west Brittany coast. Later in the year, two smaller tanker accidents, involving the *Eleni V* and *Christos Bitas*, polluted the Norfolk coast and the Welsh coast.

Increasing public awareness of the danger of oil pollution is strengthening the view that stricter action must be taken to minimise the risk of accidents. An initiative has come from the US in the aftermath of the *Argo Merchant* incident in December 1976. This ship ran aground off Nantucket Island and spilled 25,000 tonnes of oil in an area important for its fishing grounds and close to a major recreational zone.[93] The *Argo Merchant* was a 'floating rust bucket', banned from some US ports because of its pollution record, with a badly trained crew and flying under a Liberian 'flag of convenience'. The US Government intends to make vessels safer by introducing proposals for all tankers of over 20,000 tonnes to install back-up radar facilities and emergency steering equipment.[94]

Grove defends superships because most accidents involve tankers of less than 200,000 tonnes.[95] He disagrees with the view that 'flags of convenience' ships are more unsafe than others, pointing out that Liberian registered ships must have a higher incidence of accidents because one quarter of the tankers and one third of the world's tonnage are registered in Liberia. The popularity of Liberian status is due to its crew flexibility and income tax system. Nevertheless, tankers operated by ever changing crews could lead to accidents because it takes time for a crew to know their ship. Around 80 per cent of accidents at sea are caused by human error.[96] Consequently, improvements in vessel safety may be neglected by crews which work over-long hours and, in some cases, are inadequately trained.

One solution to this problem might be to implement a traffic control system for shipping similar to the successful measures adopted by aviation authorities. Ships would be guided by a port control system and would be manned by a crew operating a shift system. In Japan, a tanker entering an inland sea is under the control of the harbour authority which manoeuvres it to port with the aid of tugs. Unfortunately, captains persist in breaking the rules of the sea. The English Channel, the busiest sea lane in the world, is plagued with 'rogue' ships which cross navigational lanes to save time and money in delivery of their oil.[97] The British Government fines British ships, but only notifies the registration country of foreign offenders. After the *Amoco Cadiz* disaster, the French have adopted a tougher action by using gunboats to patrol a 12-mile limit around their coasts.

Obviously with this in mind, Shetlanders are concerned about a possible giant oilspill around their islands during the shipment of oil from Sullom Voe. Clearly, Shetland will require more assistance than is currently available to prepare adequately for a potential tanker accident.[98] The technology for cleaning up oil spills is not nearly as advanced as the technology for transporting oil. The 2.5 million gallons of detergent used to clean up the slick from the *Torrey Canyon* in 1967 still leaves traces on the Cornish coastline.[99] BP 1002, the main chemical detergent used, has been withdrawn from the market and dispersants 30 times less toxic are now in use. It is generally accepted that if an oil spill seems unlikely to cause coastal pollution, it should be allowed to disperse and degrade naturally. In Norway, methods are being developed to skim oil off the surface mechanically, separating the oil from water on board ship. The waters of the northern North Sea may be too rough to make this operation successful there, but oil separation is preferable to chemical dispersion. If the technology was developed, it would resolve the conflict between various groups concerning the use of dispersants: fishermen do not want the oil dispersed below the surface where it can be a threat to pelagic fish, and the Royal Society for the Protection of Birds would prefer the oil to be removed from the surface.

What preventive measures can be taken to avoid the necessity of cleaning up oil spills? Although the maximum fines for oil pollution in Britain at £50,000 are the severest in the world, the British courts seem reluctant to use their full powers; in 1974, the 57 convictions recorded realised a sum of only £40,000 under the Prevention of Oil Pollution Act 1971.[100] Tougher action is necessary if careless navigating is to be discouraged. The devastation of the Brittany coast by pollution from the *Amoco Cadiz* is a warning that tighter measures are required. The problems are aggravated by the inadequacy of existing procedures to cope with a major tanker accident and the lack of coordination between the various agencies involved. In 1978 the holed *Christos Bitas* was towed northwards in the Irish Sea to unload into another tanker even though it was within ten miles of Milford Haven, and the *Eleni V* was towed up and down the Norfolk coast for a month before appropriate action was eventually taken.

Future prospects

North Sea oil and gas give Britain an unforeseen breathing space for the development of alternative energy sources for the next century when supplies of hydrocarbons will begin to dwindle. Nevertheless, the reserves will not be large enough to solve all of Britain's economic problems. Britain feels that the contribution is so small (3 to 5 per cent of GNP between 1980 and 1985)[101] that the revenues derived from oil and gas would be best redistributed directly to the British people. It is extremely unlikely that the Chancellor of the Exchequer could agree to this proposal. It is more likely that long-term Government objectives will dictate a growth strategy using oil revenues either as a safety net to keep the trade balance in the black or in a planned investment programme to revitalise British industry. Whatever option is chosen Britain can take guidance from the experience of the Netherlands in its use or abuse of gas revenues. The sudden infusion of gas revenues into the Netherlands economy has contributed to the 'Dutch disease' – overgenerous social benefits, high inflation, low productivity and a collapse in investment.[102] With gas production now beginning to level out in the Netherlands, the country could have grave economic problems in the 1980s.

The size of the revenues to be channelled through the British economy will depend on the rate at which the reserves are exploited. Gas supplies will remain constant throughout the 1980s and the British Gas Corporation's main problem will be to equalise demand with supply. Oil supplies may be more erratic in the 1980s and levels of production will be related to the Government's depletion and fiscal policies. The BNOC's main role in the 1980s will be in the field of exploration as the oil majors lose interest in marginal basins; however, through participation agreements the State oil company will also play an increasingly important part in the development of new fields.

The demand for offshore supplies will begin to decline in the 1980s, and this will have repercussions on oil-related employment in the regions of Scotland As North Sea development moves into the production stage, demand for capital equipment will fall and jobs will be lost, in particular in regions such as Strathclyde, where unemployment levels are high even now. Conversely, the Grampian region, which has benefited most from oil-related employment, will consolidate its position still further in the 1980s because it is best placed to capture the offshore maintenance market. Strathclyde and other central belt regions need new investment – and may possibly attract 'downstream activities' – to replace traditional industries in order to secure a better long-term future.

The Government could also channel some of its windfall gains into tightening up pollution control measures. The *Eleni V* and *Christos Bitas* affairs highlight the inadequacies in the present system of treating oil spills, and concern has been expressed at the measures envisaged to handle a major blow-out. Prevention is always better than the cure, and a rigorous inspection system of offshore installations and tanker vessels could minimise the risks involved. Human error is often responsible for major accidents and the best the Government can do in this case is to enforce stricter penalties on 'rogue' captains and shipping companies which break the laws of the sea. International acceptance of higher safety standards at sea is imperative to minimise the present, unacceptable, high levels of oil pollution.

Notes and references

1. J. Fernie (1977) 'The development of North Sea oil and gas resources', *Scottish Geographical Magazine*, **93**, April 1977, p. 21.
2. M. Saeter and I. Smart (eds) (1975) *The Political Implications of North Sea Oil and Gas*, IPC Science and Technology Press Ltd, Ch. VII.
3. J. Grant (1976) *Independence and Devolution: the Legal Implications for Scotland*, W. Green.
4. *Financial Times*, 10 February 1978.
5. Department of Energy (1978) *Development of the Oil and Gas Resources in the United Kingdom*, Department of Energy, p. 4.
6. *Financial Times*, 25 March 1977.
7. Professor P. Odell quoted in *Financial Times*, 11 November 1977.
8. *Financial Times*, 31 March 1978.
9. P. Odell and K. Rosing (1976) *Optimal Development of the North Sea's Oil Fields*, Kogan Page; and in *Energy Policy*, December 1977, a reply to a critical book review by Kemp in the June edition.
10. P. Odell and K. Rosing (1974) 'The North Sea oil province: a simulation model of development', *Energy Policy*, December 1974.

11. D. MacKay (1978) 'North Sea oil – past lessons and future prospects', *Chemistry and Industry*, 5 August 1978, p. 545.

12. K. Chapman (1976) *North Sea Oil and Gas*, David and Charles, pp. 187, 188. See also Saeter and Smart (eds) (1975) op. cit., the discussion at the end of Ch. II, and *Financial Times*, 25 March 1977.

13. *Energy Policy*, December 1977 and June 1978, has articles by both parties on the issue of reserves. See also C. Wall and R. Dawe (1977) 'Cold water on North Sea oil reserves', *New Scientist*, 17 February 1977, pp. 407, 408.

14. A. Kemp reviewed Odell and Rosing's book in *Energy Policy*, June 1977, pp. 172–74.

15. P. Odell quoted in *Sunday Times*, 9 February 1975.

16. MacKay (1978) op. cit.

17. P. E. Kent (1975) 'Review of North Sea basin development', *Journal of the Geological Society*, **131**, Part 5, September 1975, pp. 462, 463.

18. Chapman (1976) op. cit., pp. 78, 79.

19. *Financial Times*, 31 March 1978, and personal communication to the BNOC, 16 November 1978.

20. *Fuel Policy*, Cmnd 3438, p. 7.

21. *Financial Times*, 27 May 1977.

22. *Financial Times*, 7 October 1977.

23. *Sunday Times*, 15 October 1978.

24. *Energy World*, August/September 1978, p. 24.

25. *Financial Times*, 9 November 1977.

26. Ibid.

27. *Financial Times*, April 1976.

28. Department of Energy (1976a) *Gas Gathering Pipeline Systems in the North Sea*, Department of Energy, p. 2.

29. Department of Energy (1978b) *Gas Gathering Pipeline Systems in the North Sea*, *Energy Paper No. 30*, Department of Energy; and J. Stansell (1978) 'North Sea gas – an everchanging pipedream', *New Scientist*, 27 July 1978, p. 265.

30. Buchanan and Clacher (1977) *The Collection and Disposition of North Sea Gas Liquids*.

31. *Financial Times*, 9 December 1977.

32. *Energy World*, August/September 1978, p. 24.

33. Department of Energy (1976b) *Development of the Oil and Gas Resources of the United Kingdom, 1976*, Department of Energy, p. 2.

34. MacKay (1978) op. cit., p. 546.

35. D. MacKay and G. A. MacKay (1975) *The Political Economy of North Sea Oil*, Robertson, p. 69.

36. *Sunday Times*, 4 December 1977.

37. Ibid.

38. E. Brown, Ch. V in Saeter and Smart (eds) (1975), op. cit., p. 121.

39. *Financial Times*, 2 September 1977.

40. *Financial Times*, 6 January 1978.

41. *Financial Times*, 27 February 1976.

42. Chapman (1976), op. cit., p. 88.

43. *The Observer*, 15 October 1978.

44. *UK Offshore Oil and Gas Policy*, Cmnd 5696, p. 2.

45. Saeter and Smart (eds) (1975) op. cit., Ch. IV, p. 101.

46. MacKay (1978) op. cit., p. 547.

47. S. Cochrane and J. Francis (1977) 'Offshore petroleum resources: a review of UK policy', *Energy Policy*, March 1977, p. 52.

48. *The Sunday Observer*, 23 July 1978.

49. Chapman (1976) op. cit., p. 94.

50. Cochrane and Francis (1977) op. cit., p. 58.

51. British National Oil Corporation. *Reports and Accounts, 1977*, BNOC, 1978, p. 5.

52. Ibid.

53. See Cochrane and Francis (1977) op. cit.; and Department of Energy (1978b) op. cit., p. 53.

54. Department of Energy (1976c) *Offshore Supplies Office*, OSO, p. 7.

55. Department of Energy (1974) *A Strategy for Oil Platforms*, Department of Energy, August.

56. *Financial Times*, 27 September 1977.

57. A £2-million contract for a 2,600-tonne wellhead platform for the Fulmar field and a £4-million contract for a 2,400-tonne deck for the Beatrice field.

58. Fernie (1977) op. cit. p. 26.
59. Ibid.
60. *Financial Times* survey, 'Offshore maintenance', 20 April 1977.
61. Department of Energy (1977) *Development of the Oil and Gas Resources in the United Kingdom*, Department of Energy p. 10.
62. MacKay and MacKay (1975) op. cit., p. 133.
63. Of the 37 million tonnes of oil produced in 1977, 21 million were refined in the UK.
64. *The Economist*, 11 February 1978, p. 89.
65. Scottish Council (Development and Industry) (1977) *Petrochemicals in Western Europe: the Potential for North Sea Oil and Gas*, SCDI; and *Financial Times*, 4 May 1977.
66. *Financial Times*, 19 May 1977.
67. Scottish Development Department (1974) *North Sea Oil and Gas: Coastal Planning Guidelines*, HMSO.
68. M. Hill (1976) *Oil Over Troubled Waters*, Aberdeen People's Press.
69. D. W. Taylor (1974) 'Offshore oil: a cause for regret?', *Architect's Journal*, 26 June 1974.
70. J. Francis and N. Swan (1974), in *Scotland's Pipedream* (St Andrew's Press), have also outlined the social problems which can occur due to the influx of a large temporary workforce to an area like Peterhead.
71. *Financial Times*, 14 January 1977.
72. E. Balneaves (1978) 'Fortunes of the Shetlands', *Geographical Magazine*, August 1978.
73. Sumburgh handles more passengers per year than Prestwick, Scotland's international airport: Balneaves (1978) op. cit., p. 729.
74. Ibid.
75. The average wage is £250 per week. 'News at Ten' reported, on 24 November 1978, that a couple working at the terminal intended to *save* £33,000 in three years.
76. M. Kenward (1978) 'Oil companies nurture the golden goose', *New Scientist*, 28 September 1978, p. 928.
77. Department of Energy (1978b) op. cit., p. 49.
78. C. Milner (1978) 'Shetland ecology surveyed', *Geographical Magazine*, August 1978, p. 730.
79. C. Skrebowski (1978) 'Is Britain's oil industry on shaky legs?', *New Scientist*, 16 March 1978, p. 714.
80. *Financial Times*, 20 April 1977.
81. *Financial Times*, 28 January 1977.
82. Balneaves (1978) op. cit., p. 726; and J. Button (ed.) (1976) *The Shetland Way of Oil*, Thuleprint, p. 102.
83. 'Disaster Diary', part of the BBC series, *The Energy File*, 6 April 1977.
84. Department of the Environment (1976) *Accidental Oil Pollution of the Sea, Pollution Paper No. 8,* Department of the Environment.
85. *Financial Times*, 24 June 1977.
86. Under an international convention of May 1977, compensation will be payable by offshore operators for oil pollution caused by their activities in coastal waters of countries which join the Convention. The maximum amount per incident for compensation will be $35 million rising to $45 million in 1982.
87. *Financial Times*, 16 November 1978.
88. L. Johnston in Button (ed.) (1976) op. cit., p. 65.
89. A quote attributed to J. Fay (1978) in *Planning* No. 263, 14 April.
90. BBC programme, *Brass Tacks – an area of outstanding danger*, 13 September 1978.
91. Fay (1978) op. cit.
92. N. Grove (1978) 'Giants that move the world's oil: Superships', *National Geographic Magazine*, **154**(1) July 1978.
93. For a detailed account of the *Argo Merchant* disaster see the *Observer Magazine*, 24 April 1977.
94. *Financial Times*, 25 May 1977.
95. Grove (1978) op. cit.
96. BBC programme, *Man Alive* – 'Standing into danger', 14 November 1978.
97. Ibid.
98. Johnston in Button (ed.) (1976) op. cit.
99. *Sunday Times*, 18 May 1975.
100. Ibid.
101. *Financial Times*, 26 May 1977.
102. *Sunday Times*, 4 December 1977.

Chapter 4

Nuclear power: salvation or damnation?

The future of nuclear power in the UK aroused public interest with the publication of the Flowers Report[1] in 1976, the Windscale public inquiry in 1977 and the prospect of a fast breeder reactor inquiry in 1980. The nuclear power issue had been relatively underemphasised in the 1970s perhaps because the 'energy crisis' and North Sea oil expectations diverted attention elsewhere. In other countries, however, public opposition has caused a slowdown of reactor construction, notably in the USA, Italy and West Germany. Denmark, with a lack of indigenous energy resources, has rejected the nuclear option; the Social Democrats in Sweden were ousted from power by an opposition party committed to reviewing nuclear policy.

Decisions to develop or curtail nuclear power are essentially political and in Britain Mr Tony Benn, when Secretary of State for Energy, adopted a policy of open government in order to keep as many options open as possible. The task is not an easy one. The Select Committee on Science and Technology was highly critical of the machinery for nuclear decision making when it assessed the Steam Generating Heavy Water Reactor (SGHWR) programme, commenting that 'after extensive private and public debate sufficient information is apparently still not available for the country to proceed with confidence – at whatever pace – to the construction of new nuclear power stations.'[2] The Secretary of State has declared that if the public is to accept nuclear power, nuclear affairs should not be veiled in secrecy and more information should be made public.[3] At the same time, the public began to demand more nuclear knowledge and after much disquiet about British Nuclear Fuels Limited (BNFL)'s expansion plans at Windscale, Mr Shore, the Environment Secretary, ordered a public inquiry. The Government's adoption of the Town and Country Planning Association's suggestion that an independent standing commission should be created to advise on energy policy and the environment may enable the participants in future inquiries to argue from a more informed base.[4] Do we need nuclear power? If the answer is affirmative, at what pace should the nuclear programme develop? If the answer is negative, for what reasons should we contemplate abandoning a programme that absorbs the greatest allocation of the Government's expenditure on energy research and development. The outcome depends on the scale of the impending 'energy gap' referred to in Chapter 1 and on the resolution of the apparently irreconcilable views that, on the one hand, nuclear power will be our energy salvation in the twenty-first century as fossil fuels are diverted to non-electricity markets while, on the other hand, a nuclear future will inevitably lead to nuclear accidents, proliferation of weapons, an infringement of personal liberties, comprehensive damnation.

This chapter will outline the role of nuclear power in overall electricity planning. In 1977 nuclear power provided only 13 per cent of the nation's electricity; by 2000 this figure could be over 70 per cent. By assessing past and present forecasting of electricity demand, an attempt will be made to substantiate or repudiate such a forecast. In the second half of the chapter, the environmental aspects of nuclear power will be examined to test the Atomic Energy Authority Chairman's claim that this form of energy is safer, cleaner and has less impact on the environment than any practical alternative way of producing power for the nation's needs.[5]

The electricity industry

The electricity supply industry was nationalised in 1948 to incorporate 560 undertakings which produced electricity throughout the UK. From this date until the recession of the middle 1970s, the supply of electricity increased at an average annual rate of 7 per cent, 5 per cent higher per annum than the average figure for all energy supply industries combined.[6]

The types of fuel used in power stations have changed according to economic considerations throughout the period. On nationalisation, almost all of the electricity generated was from coal-fired stations; by 1976, 76.6 million tons of coal were burned but the tons of coal equivalent for oil, natural gas and nuclear generation were 17 million, 2.6 million and 12.7 million, respectively. The development of nuclear power and the increased burning of oil in power stations were policies initiated by the Government in the 1950s to preserve coal for non-electricity markets during a period of energy shortage. By the 1960s the availability and relative cheapness of oil left coal at a competitive disadvantage. Undoubtedly more power stations would have been constructed or converted to oil burning if the Government had not introduced the Coal Industry Act in 1967 to subsidise an increase in coal burning to protect the coal industry. OPEC price increases in the 1970s have redressed the price differential between oil and coal: nevertheless, with the long term prospect of a widening energy gap, fossil fuels may increasingly be diverted to more efficient, premium uses with nuclear power supplying the greatest part of the electricity market.

The location of power stations

The criteria governing the siting of stations burning fossil fuel differ from those for nuclear stations. When the British Electricity Authority inherited hundreds of coal-burning stations in the 1940s, most towns generated their own electricity and coal-deficient areas imported coal to fire their plants. With the increase in size of generating sets and the development of a super grid network, areas close to fuel supplies, water provision and the grid became favoured locations. Gradually, the geographical pattern of power stations illustrated in Fig. 4.1 evolved. The Yorkshire and East Midlands coalfields met the locational criteria and many new power stations, such as Ferrybridge, West Burton and Drakelow, were commissioned in these areas in the 1960s. During the same period, the Cockenzie and Longannet plants came into operation to serve the central Scotland market, utilising coal from the Lothian and Fife coalfields. Longannet, commissioned in 1970, is the largest power station in Britain with four 600 MW sets. Oil-fired plants are located near refineries, which sell the fuel 'over the fence' for burning. Although Fawley and Pembroke consume imported oil, the Peterhead station, under construction, is dual-fired to burn North Sea oil and gas.

Fig. 4.1 Location of power stations (*Source:* Central Office of Information [adapted] *Energy,* HMSO, 1977, and *CEGB Statistical Yearbook* 1977–78)

The location criteria for nuclear stations involve safety considerations and the type of fuel used in the reactor. Unlike coal, oil or gas, uranium fuel is light, easy to transport and can be used for several years before it is irradiated. The general situation of nuclear plants is therefore less dependent on proximity to raw materials and communications networks than that of fossil fuel power stations. The chosen site may, however, be similar in local attributes – nuclear stations still require an adequate geological foundation to support the heavy plant as well as water nearby for cooling purposes, and of course close access to the grid. The latter condition is most difficult to fulfil since nuclear plants have been located away from major centres of population because of the danger of a nuclear accident. Hence preferred nuclear sites are in remote coastal locations or, like Trawsfynydd, beside an inland lake. However, the second generation of reactors, the Advanced Gas Cooled Reactors (AGRs), are enclosed in a prestressed concrete vessel and are regarded as safer. This led the Ministry of Power to modify its policy in 1968 to allow siting close to built-up areas.[7] The AGRs under construction at Heysham and Hartlepool are results of this policy.

The geography of nuclear energy is not concerned solely with power production. This is only one part of an intricate cycle which involves the mining, enriching and fabricating of nuclear fuel prior to its use in the reactor (Fig. 4.2). When the fuel becomes irradiated, it can be reprocessed for re-use to complete the cycle. With the exception of the mining itself, which is concentrated in the USA, Australia, South Africa and Canada, the UK nuclear industry is one of the world's leaders in all aspects of the fuel cycle. Britain, for example, has over a quarter of a century of experience in the reprocessing of irradiated fuel. In order to illustrate the processes involved in the fuel cycle, an account of the development of nuclear energy is necessary.

The early years in the nuclear industry

The early development of nuclear power was for military rather than civilian purposes. The end of the Second World War after the destruction of Hiroshima and Nagasaki by nuclear bombing led to an end of the cooperation between the three countries involved in the Manhattan Project for nuclear research and development. The McMahon Act of 1946 made it illegal for Americans to divulge nuclear information to their former allies; the Canadians decided to embark on a civilian nuclear programme, whilst Britain channelled her efforts into a nuclear weapons programme. Research and development concentrated on improving the efficiency of weapons and a move from fission (A-bomb) technology to fusion (H-bomb) technology. Indeed, public opposition to nuclear power in the 1950s and 1960s centred on 'Ban the Bomb' campaigns and the number of nuclear tests, which reached a peak of 133 in 1962.[8]

The foundations of the British nuclear industry were laid in the immediate post-war years.[9] Although this was a period of material shortages and financial restrictions, the infant industry, under the leadership of Christopher Hilton, began impressively. By 1952 the first four nuclear factories had been commissioned. They were:

(a) the uranium refinement and fuel fabrication factory at Springfields, near Preston;

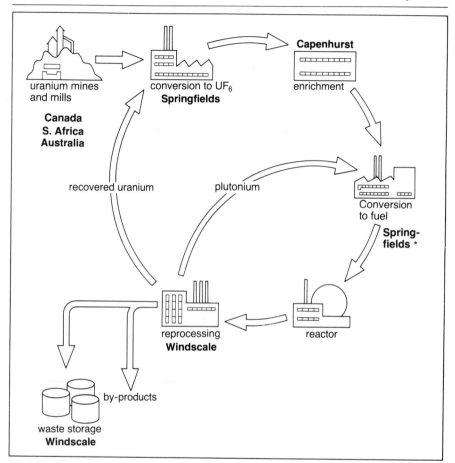

* Manufacture of plutonium fuel is carried out at Windscale

Fig. 4.2 The nuclear fuel cycle (*Source:* W. Patterson: *Nuclear Power,* Penguin 1976)

(b) a plutonium production plant; and
(c) a reprocessing plant at Windscale;
(d) an enrichment plant at Capenhurst, Cheshire.

Within a few years, the plutonium production reactors at Windscale were augmented by further reactors at Calder Hall and Chapelcross (Fig. 4.1). The first reactor at Calder Hall has the distinction of being the world's first commercial nuclear power station,[10] although the electricity which first entered the grid in 1956 was really only a by-product; the Chapelcross and Calder Hall reactors were primarily used to produce plutonium for the UK's weapons programme.

Nevertheless, the energy shortages of the early 1950s caused the Government to create the United Kingdom Atomic Energy Authority (UKAEA) in 1954 and the first civil nuclear power programme was announced in the White Paper of the following year. The anticipated 'energy gap' did not materialise; oil became cost-competitive and nuclear costs were so high that the original programme to commission 12 stations was modified by the Government – eventually only nine Magnox power stations were constructed. In 1962 the first two plants at Berkeley and Bradwell were commissioned and by 1971 the final station, Wylfa, came to full power to mark the end of Britain's first generation of civil power reactors (Table 4.1).

Table 4.1 Nuclear power programmes

Type of reactor	Date commissioned	Location
Magnox	1962	Berkeley
	1962	Bradwell
	1965	Hinkley Point 'A'
	1965	Trawsfynydd
	1965	Dungeness 'A'
	1966	Sizewell
	1967	Oldbury
	1971	Wylfa
AGRS	1976	Hinkley Point 'B'
	1976	Hunterston 'B'
	1978 (under construction)	Hartlepool
	1978 (under construction)	Heysham
	1978 (under construction)	Dungeness 'B'

Nuclear fission

The Magnox and subsequent commercial nuclear energy programmes are based on the process of nuclear fission. The 'splitting of the atom' involves the bombardment of the nucleus of the uranium isotope U-235 by stray neutrons. When U-235 undergoes fission, further neutrons are released which may strike other nuclei to produce a chain reaction; similarly if the nucleus of the heavier uranium isotope U-238 reacts with neutrons the fissile plutonium 239 is created, which also undergoes a chain reaction. As the fissile U-235 only occurs as 0.7 per cent of natural uranium, the fissile nuclei are too widely spaced to sustain fission; hence, U-235 is either enriched to increase its proportion to 2 or 3 per cent or the neutrons from fission are slowed down to enable them to be more easily absorbed by U-235. The speed of the neutrons can be moderated by surrounding the uranium fuel with a light nuclei substance. Heavy water (deuterium oxide) is the best moderator but carbon in the form of graphite is cheaper, and as a solid is easier to manipulate within a reactor. To control the reaction itself control rods absorb excess neutrons within the reactor core. The rods are then withdrawn to build up the neutron density until 'criticality' is reached and a sustained chain reaction occurs. The uranium fuel becomes hot and this heat is removed by pumping a cooling fluid through the core of the reactor. Originally the heat released was superfluous to the aim of producing plutonium for weapons, but civil programmes use the heat to produce steam to drive electricity generators.

The uranium metal fuel used in Magnox reactors is clad in magnesium alloy, a graphite moderator slows down neutron speed and the coolant is carbon dioxide. The problems in the British nuclear industry began in the 1960s when the Government had to choose a reactor design to follow the Magnox. In Britain, the logical development of the Magnox reactor – the Advanced Gas Cooled Reactor (AGR) – was being tested with the 32 MW Windscale prototype. The smaller, more compact design would improve the fuel output per kilogram of uranium, minimise the frequency of refuelling and reduce the high capital costs incurred in the construction of Magnox stations. The uranium metal fuel could not withstand high temperatures for long periods, and so the AGRs use uranium oxide, which can withstand more than twice the temperature of natural uranium. This fuel is clad in stainless steel – an absorber of neutrons – and requires enrichment. The competing American designs – the Pressurised Water Reactor (PWR) and Boiling Water Reactor (BWR) – are light water reactors. Both designs use water as a moderator and a coolant, but water absorbs sufficient neutrons for the fuel to need to be enriched.

In May 1965 the Government announced the second nuclear programme based on the AGR design. Henderson claims that this programme was a major error, comparable with Concorde as a financial disaster. Technical difficulties were greater than expected with AGRs and the delays in construction led to an escalation in costs.[11] When the Hinkley Point B plant was completed in 1976, construction costs totalled £138 million, 45 per cent higher than the figure negotiated with the contractor in 1967.[12] Only two AGRs were in operation by 1976 but both were shut down in 1977 – the Hunterston B station for emergency repairs when seawater entered the reactor. The problems encountered with the AGR design have allowed the American PWR reactors to consolidate their hold on world markets.

Enrichment

Unlike the Magnox and the Canadian-designed Candu reactors, which use natural uranium as fuel, the new generation of reactors, the AGRs, PWRs and BWRs, uses uranium oxide which, as we have seen, requires enrichment to increase the proportion of uranium 235 nuclei in the material to between 2 and 3 per cent, depending on reactor design. The earliest enrichment plants at Oak Ridge, US, Capenhurst, Cheshire and Pierrelatte in France were built to enrich uranium for nuclear weapons. The construction of two further plants in the US increased American output to 17,230 tonnes per year in the 1960s;[13] this made the United States the principal supplier of enriched uranium to the West. The US dominance of the Western market was broken by the Soviet Union in the 1970s as it began to supply enriched fuel to West European countries, including Britain. At the same time, these West European consumers began to develop their own enrichment technologies. The US and USSR both use the energy-intensive gas diffusion technology. The Dutch, British and German combine Urenco Centec has prototype plants at Capenhurst and Almelo (Netherlands) which are developing the new gas centrifuge technology in the hope of reducing initial energy inputs by one-tenth.

Policy in the 1970s

Although the British nuclear industry was developing other aspects of the fuel cycle, a decision had to be taken about the choice of reactor for future nuclear power programmes. The Government had to decide whether to persevere with the troubled AGR design, to license PWRs for construction in the UK or to scale-up the prototype Steam Generating Heavy Water Reactor (SGHWR) which had been in operation since 1967 at Winfirth.[14] Although the CEGB under Sir Arthur Hawkins favoured the PWR design, the Government announced in July 1974 that six SGHWRs with a total capacity of 4,000 MW would form the third nuclear power programme. The safety of PWRs had been seriously questioned by leading scientists and this may have influenced the decision. It certainly prompted the Atomic Energy Authority to conduct a safety study of PWRs and in 1976 it reported that the design could be licensed to meet the standards of the Nuclear Inspectorate.[15] Meanwhile, the SGHWR programme was behind schedule and the first AGRs had begun to supply power to the grid. Eventually, the SGHWR programme was dropped in favour of the other reactor options. Two new AGRs were ordered in 1978 and a design study costing £40 million was initiated on the PWRs.[16]

The problems of reactor choice have been further complicated by the slump in electricity demand and the economic ill-health of the power plant manufacturing industry. The Central Policy Review Staff (the 'Think Tank') condemned the irregular nature of CEGB power-plant ordering and proposed that a forward ordering programme of 2,000 MW per annum should be implemented to save the power-plant industry from collapse until demand picked up in the 1980s.[17] The Government's response to this plea was to subsidise the CEGB for the advanced ordering of the Drax B *coal-fired* station in 1977. The ramifications of this policy will be discussed at greater length later in the chapter, but it is obvious that the weakest competitive aspect of the fuel cycle for Britain is now the once glamorous technology of the reactor. One door that still remains open to Britain and other industrialised countries which seek markets for the reactors is the export of Fast Breeder Reactor (FBR) technology. Since the US has deferred a decision on reprocessing and its FBR programme, other countries, such as Britain and France, could capture the major share of the potential export market.

Reprocessing and the FBR

Although the Windscale inquiry and the impending fast breeder reactor inquiry are being classed as separate issues, the plans for both programmes are fundamentally linked since the reprocessed plutonium which is recovered from the depleted fuel can be then used in the FBR. Britain has unparalleled experience in reprocessing technology: the Windscale plant was commissioned in 1951 and in 1964 a new plant, B205, was constructed to handle the oxide fuel from the new generation of reactors. The original aim of the nuclear industry was to separate plutonium from the irradiated fuel of military reactors, but energy shortages in the 1950s and a lack of indigenous uranium stimulated a Research and Development programme into FBR technology. The demonstration FBR at Dounreay was producing power as early as 1959. The inherent advantage of FBRs is that they 'breed' fuel; the present generation of reactors are 'burner' reactors because they create less fissile material than the amount of original uranium used. Potentially, FBR technology allows uranium to be burnt 50 to 60 times more efficiently than in burner reactors, but the engineering difficulties have yet to be resolved. The fundamental design problem is related to the geometry of the core of the reactor. As mentioned earlier, burner reactors require a moderator to slow down thermal neutrons to create fission; it takes 400 times as many *fast* neutrons as thermal neutrons to cause fission in FBRs. Consequently, to achieve a high neutron density, a compact core has to be designed without a moderator but with a coolant which can quickly remove the heat.[18]

The US, which introduced the first experimental FBR in 1951, has slowed down its present programme after a series of mishaps in the 1960s and the more recent Ford/Carter initiatives on reprocessing. At present, Britain, along with France and the USSR, have the only prototype FBRs. Adjacent to the new 250 MW FBR at Dounreay is a reprocessing plant which recovers the newly 'bred' plutonium. This technology closes the fuel cycle.

Originally the UK Atomic Energy Authority hoped to introduce commercial size demonstration FBRs in the 1970s, but design modifications to meet stringent safety standards have reduced its breeding efficiency and increased costs. It now seems unlikely that the FBRs will be able to provide significant amounts of electricity by the end of the century. The arguments for a delay or postponement of a commitment to FBRs have stressed the seeming irrelevance of this technology to the major energy problems of the country in addition to the environmental problems involved.[19] These issues came dramatically into the public eye during the Windscale inquiry.

In 1974 British Nuclear Fuels Limited (BNFL) made known its plans to build a new Thermal Oxide Reprocessing Plant (THORP) at Windscale in Cumbria. The formal application was submitted in June 1976, but in addition to the THORP proposal BNFL also wished to refurbish their existing Magnox plant, construct a £40 million plant to perfect the glassification of nuclear waste and other minor work such as the provision of more storage ponds. Pressure for a public inquiry mounted as the Flowers Report indicated that a long-term commitment to nuclear power was undesirable until the storage of high level waste had been proven safe beyond all reasonable doubt.[20] In the same year, President Ford had stated that reprocessing was not an inevitable step in the nuclear fuel cycle. Eventually the Environment Secretary, Peter Shore, announced that a public inquiry would take place in June 1977 on the condition that BNFL submitted its planning application in separate parts in order that only the THORP proposal would come under public scrutiny.

During the 100 days of the inquiry most of the arguments raised were centred on the environmental impact of the reprocessing plant, and these will be discussed in more detail later in the chapter. The main point that concerned the scheme's opponents, led by the Friends of the Earth (FOE), was the safety of disposal procedures for the waste from the plant. FOE argued that evidence from North America had shown that since oxide fuel could be stored for at least 12 years the decision on reprocessing could be postponed until vitrification technology for safe disposal had been perfected.[21] Alternatively, it was argued, the scale and potential dangers of the project could be lessened by reopening the existing oxide fuel reprocessor on the site. This course would cost only £5 million, but would have sufficient capacity to handle irradiated fuel from British reactors as against the proposed £600 million plant which would also reprocess other countries' fuel.[22] At the end of the inquiry Mr Justice Parker recommended that THORP be given planning permission without delay. He dismissed the possibility of prolonged storage as too costly because it would require the development of new storage methods.[23]

However, Peter Shore rejected the application and re-introduced it under a Standing Development Order to allow the issue to be debated in Parliament. In May 1978 the House of Commons debated the proposal and, although the Liberals attempted to block the plan, approval was granted by 224 votes to 80.

The Windscale decision means that plutonium will be available in sufficient quantities to develop a FBR programme. Nevertheless, the Government continues to explore all the other options available to it in the fuel cycle. A commitment to nuclear expansion could leave Britain open to OPEC type action from suppliers. Fortunately, most of the uranium suppliers are industrialised countries which have good diplomatic relations with Britain although even they have been keeping a tighter rein on their supplies in recent years.

The nuclear fuel cycle is becoming an intriguing political battle. The US and Canada's rejection of reprocessing could mean that other countries such as Japan which depend on US enriched fuel, become forced into storing their waste rather than reprocessing it in the interests of 'nuclear non-proliferation'. Already, the Fox Report's main objection to an expansion of uranium mining in Australia concerned the dangers of proliferation.[24] Canada has adopted a more cautious policy in its exporting of nuclear materials since 1974 when India developed a nuclear weapon from an imported Canadian reactor intended for power generation.

In the United States, President Carter has not only backed the non-proliferation initiative but has also supported a report from the Nuclear Energy Policy Study Group which claims that the abundance of world uranium reserves will make thermal reactors more competitive than FBRs for longer than originally anticipated.[25] Hence, with uranium prices unlikely to exceed $30 to $40 per lb even by the year 2000, the US feels that the development of FBRs and reprocessing is not a matter of urgency. The US Government was unwilling to endorse the Windscale proposals and was concerned at the apparent misunderstanding of US policy which arose at the inquiry. In February 1978, the Americans requested, through diplomatic channels, that the British Foreign Secretary clarify the position to the Inspector by stating their opposition to the planned construction of the reprocessing plant, mainly on the grounds of nuclear proliferation.[26]

The British Government is in a dilemma which is reflected in the delays in decisions concerning the FBR and other reactor designs. The delicately balanced political situation worldwide could restrict uranium supplies to the UK: reprocessing could be too dependent on political developments outside Britain's control. The South of Scotland Electricity Board (SSEB) obtains uranium from Canada and South Africa for its two nuclear stations. Perhaps fearing political circumstances which could restrict supplies, the SSEB has been exploring the possibilities of mining uranium in Orkney. Previous studies of the 'uranium corridor' from Yesnaby to Stromness had suggested commercial mining possibilities, and the SSEB plans to drill 11 test boreholes in this area. The secrecy with which the negotiations were handled has provoked much antagonism from the Orcadians towards the SSEB. Most Orcadians knew nothing of the scheme until the SSEB presented the Council with a *fait accompli*. To the disgust of the National Farmers' Union, the Board had made agreements with 40 farmers prior to submitting a planning application.[27] The

SSEB plans will meet much opposition from the Orcadians if a public inquiry is held.

Nuclear fusion

The ultimate solution to energy shortages in the twenty-first century could lie in the research and development presently being carried out into nuclear fusion. Fusion technology attempts to simulate the thermo-nuclear reactions responsible for the sun's output of energy. A controlled H-bomb explosion is not technologically impossible but such a vast amount of power is required to heat the nuclei that at least at the present level of technology, the process consumes more power than it delivers.

In the US, USSR, Japan and the EEC, fusion scientists are experimenting with thermo-nuclear devices to work toward a fusion reactor design. Research and development expenditure is high and costs are escalating. While the governments of the EEC wrangled over which of them was to be the host country for the ambitious Joint European Torus (JET) scheme, the estimated cost rose from £70 million to £119 million between February 1976 and March 1977.[28]

What are the attractions of a scheme which incurs high costs and cannot guarantee to produce commercial electricity in the foreseeable future? The key to its lure is that it involves the fusion of two hydrogen isotopes which are abundantly available compared to uranium. These are deuterium, which occurs in ordinary water, and tritium, which although it does not occur naturally, but is created by the neutron bombardment of lithium, is potentially four times more abundant than uranium.[29]

JET is essentially a giant electo-magnet shaped like a doughnut. For fusion to occur, the deuterium and tritium need to be heated at extreme temperatures and pressures; the hydrogen isotopes lose their electrons and the 'plasma' created is contained within a magnetic field. The plasma has to be held together at high temperatures for fusion to occur; if these conditions are not realised either fusion does not take place or the amount of energy generated is less than that required to sustain the reaction. The JET experiment will operate at power levels as high as 230 MW.

The demand for electricity

Forecasting demand poses a difficult problem for the individual electricity boards and the Electricity Council. The accuracy of forecasting is greatly lessened by the length of time needed to plan and construct new power stations. The example of the AGRs neatly illustrates this. If demand is underestimated, facilities become stretched; if demand is overestimated, expensive plant could be left idle for much of the year. The current problems of surplus generating capacity are mainly a legacy from decisions taken in the 1960s. The growth of electricity demand in the 1950s took the CEGB by surprise and after the harsh winter of 1963, which had plant capacity fully stretched, more orders for additional capacity were placed in the 1960s. The steady expansion predicted for electricity demand in the 1970s has not materialised; as late as 1973 the CEGB intended to build 32 1300 MW PWRs within a decade, anticipating an increase in demand of 5 per cent per annum. This plan was of course superseded by the SGHWR programme but by 1975 it became clear that the CEGB must reconsider its policy. Demand had not increased; in fact it fell in 1974 and 1975. The CEGB therefore decided to curtail power station development.

Since the 'energy crisis' and the Government's policy of phasing out subsidies to the nationalised fuel industries, the price of electricity has risen faster than that of its main competitor, natural gas. The result has been a fall in demand for electricity with the supply system operating at only 70 per cent capacity. In 1975/76 maximum demand on the electricity supply system was 46.7 GW from a generating output capacity of 66.9 GW.[30] Allowing for the 20 per cent plant margin allowed for breakdowns and maintenance, 10.9 GW of capacity were superfluous to requirements. Despite this surplus capacity, the Energy Commission envisages a build-up of additional capacity of 10 to 25 GW by 1990 but even the lower estimate must be considered optimistic. If the Commission's assumption that demand will increase by 3 per cent per annum is accurate, by 1980/81 maximum electricity demand will be 54 GW; however, output capacity will have increased to 83.5 GW because of plant already under construction (this does *not* include Drax B or the two new AGRs which have been ordered). Surplus capacity will increase and it is unlikely that a new order for a power station will be necessary until the early 1980s – mainly to replace obsolete plant.

The CEGB raised the plant margin to 28 per cent in January 1978, which may mean either that the large modern plant is less reliable – a possibility considering the shaft failures in two 600 MW generators at Drax A and the technical problems with the only two AGRs in operation – or that the CEGB is attempting to conceal the lack of demand for electricity by 'inventing' another 8 per cent of plant margin capacity. The CEGB is endeavouring to recapture lost demand with promotional campaigns such as 'Build Electric', extolling the virtues of insulation in all-electric Medallion homes. As Robert Vale points out, the Medallion home is in fact not very successful in promoting energy conservation.[31] If large numbers of these houses were built, national electricity consumption would certainly increase because of electricity's low conversion efficiency for space heating purposes. Vale cynically comments that homes like these designed without flues and with electricity as their sole source of fuel can only benefit the electricity workers since they would make future industrial action more effective. The potential homeowners may not be so easily found, however, because independent leaflets such as the Department of Energy's 'Compare Your Home Heating Costs' show electricity at a competitive disadvantage with other fuels.[32]

The power manufacturing industry and employment

It is clear that the CEGB is in a dilemma. If it orders more plant which is not needed, either electricity prices will have to rise – further depressing demand – or the Government will have to subsidise the scheme at the taxpayers' expense. If it does not order new plant, the power manufacturing industries will have to make redundancies because of a lack of orders. Government policy has been to bolster employment in this sector; in 1972 the Ince B station was ordered early and subsidised, and similarly in 1977 the Government agreed to pay the costs of advanced ordering for Drax B. Unfortunately, the current situation is more acute: home orders are not forthcoming and the overseas market is exceedingly competitive since other countries (like the US) also have surplus capacity problems and are attempting to make up for a famine of orders. Since much of the export potential is in the nuclear field, Britain is at an added disadvantage because she has not developed a reactor to sell 'off the shelf' like the PWR. The Central Policy Review Staff's remedial action would be costly to the taxpayer. To save the 34,000 jobs in the boiler-making and turbo-generator industries, it suggested that the Government must prevent the collapse of the industry by subsidising forward ordering until electricity demand recovers and more capacity is needed in the 1980s.[33] Other recommendations include the merger of the turbo-generator groups GEC and Parsons to create a competitive unit in world markets and a move to larger generating sets to achieve economies of scale.

The Government has adopted some of these measures – the advanced ordering of Drax B and two AGRs; the question of the merger is unresolved because the parties involved cannot agree on its form. Nevertheless, at present the capturing of foreign markets can be only achieved if uneconomic tenders are submitted.

The Central Policy Review Staff's recommendations are considered by some to be a waste of public money. Are the policymakers asking the wrong questions? The Government, yielding under political pressure, has ordered stations it does not need because Babcock and Wilcox (Clydeside) and Parsons (Tyneside) are in areas of existing high unemployment. The CEGB has replied to criticism of concentrating on 660 MW sets for the export market by arguing that larger sets have no market potential overseas. Perhaps a step *backwards* is right. Patterson quotes Lucas, who claims that the optimum size of generating plant is between 200 and 250 MW.[34] Furthermore, the main reason for the 'blackouts' during October 1977 after an unofficial strike by power workers was not lack of plant but the inflexibility of large units which could not respond to changes in demand. A change of policy would be desirable and the Government could invest in *existing* plant to cure the major problem of the unreliability of large sets and improve the thermal efficiency of old plant.

The prevention of redundancies by advanced ordering causes 'unseen' problems of a loss of jobs elsewhere in the energy supply system. The CEGB, for example, maintains that the ordering of Drax B, with its better fuel-burning efficiency, will ultimately lead to a reduction of 2000 man-years in the CEGB's labour requirements as older, inefficient plants are closed.[35] In addition, coal sales would be curtailed by 1.5 million tonnes resulting in a loss of 3,000 man-years in the coal mining industry.

The short-term problems for the coal industry could be embarrassing for the Government, which has encouraged the NCB in its plans for expansion. Manners maintains that its decision to order two new nuclear stations has overlooked the implications for its coal policy.[36] The plants under construction are predominantly oil or nuclear stations; by 1980 the proportion of coal capacity to the total will fall from the present 65 per cent to 51 per cent.[37] The ordering of new nuclear plant can only aggravate the plight of the coal industry whose main customer is the CEGB. Owing to an excess of capacity, 48 small power stations were closed between October 1976 and March 1977, causing a loss of 5,000 jobs.[38] The Energy Commission summarises the expected situation by 1980 (Fig. 4.3). Although coal accounts for 65 per cent of the total number of plants,

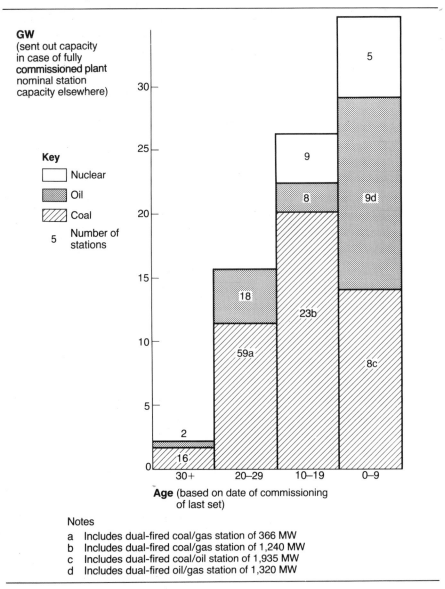

GW
(sent out capacity
in case of fully
commissioned plant
nominal station
capacity elsewhere)

Key

☐ Nuclear

▨ Oil

▨ Coal

5 Number of
   stations

Age (based on date of commissioning
of last set)

Notes

a Includes dual-fired coal/gas station of 366 MW
b Includes dual-fired coal/gas station of 1,240 MW
c Includes dual-fired coal/oil station of 1,935 MW
d Includes dual-fired oil/gas station of 1,320 MW

Fig. 4.3 Expected age distribution of UK public supply generating capacity in 1980 (*Source: Energy Commission Paper Number 1*)

most have a small generating capacity, with 74 out of the 103 plants being over 20 years old. It is possible that if demand continues to be depressed, the 16 coal power stations over 30 years old will be prematurely closed.

The key to the complex inter-relationships between the CEGB and the NCB over an ordering programme would be to freeze further orders for five years and conduct a thorough modernisation of existing plant. By keeping the older coal plants open and increasing their efficiency, less coal would be required but this measure would minimise the drastic effect a closure policy would have on the NCB's short term policy. At the same time, improving reliability in the larger sets would reduce the necessity for such a wide plant margin, thus utilising capacity more efficiently. Redundancies are inevitable in the power manufacturing industries but an emergency forward order programme is not the best way to save jobs. A steady flow of small orders would be won and fewer jobs would be lost if a modernisation programme was adopted. Patterson goes so far as to say that reliable small sets would have more export potential than the glut of large sets that are in competition in world markets.

The economics of nuclear power

The costing of nuclear power generation is fraught with difficulties because of accounting complications inherent in the fuel cycle. The costing of a fossil fuel plant is fairly straightforward; it simply involves costs of construction, fuel, transportation, plus running and maintenance costs. With nuclear stations the sums are more complicated because, in addition to construction costs and other 'on site' running costs, the cost of fabrication and enrichment (for oxide fuel) before transportation to the plant has to be taken into account. Additional costs arise in the reprocessing or storing of irradiated fuel. An important factor, often overlooked, is the cost of decommissioning the plant at the end of its active life.

Bethe, arguing a case for the necessity of fission power in the United States, points to the 'hidden investment' costs of the alternative – coal-fired power plant.[39] The cost of new mines, infrastructure and compensation to miners for lung diseases are all cited in turn; however, it is rather easier to find hidden or rather ignored costs in the nuclear industry. When the nuclear fuel reaches the power station, its quoted price does not reflect its true economic cost. The first enrichment and reprocessing plants were financed out of military budgets and hence 'written off' as a component of investment for electricity generation. Only in the 1970s, as replacement plant is being constructed, has any indication of the real costs become apparent. The US, in any case, monopolised the enrichment market by selling at subsidised prices. The real economic cost of such a plant is demonstrated by the fact that even advanced and populous countries such as Britain, the Netherlands and West Germany need to share the capital costs of construction by establishing the joint venture Urenco Centec. In 1975 the combine announced its terms for enrichment; they were twice that of its US and Soviet competitors.[40] Reprocessing plants are also uneconomic propositions. This fact is recognised by Mr Justice Parker in his Windscale Report.

His recommendation to proceed with THORP without delay may be sound advice considering the escalation in cost estimates since the late 1960s. Originally it was hoped to convert a military reprocessing plant for £3 million, and by 1976 the figure quoted for a new plant was £350 million; at the inquiry the cost estimate was £600 million. In addition to raising this capital from public funds, BNFL justifies its economic credibility by asking its customers, in the first instance the Japanese, to pay towards the cost of construction.

In the US private companies are unwilling to take the financial risk of building reprocessing plants. Getty Oil ran a plant for six years at Buffalo which lost \$7.5 million a year until it closed in 1972; similarly, General Electric lost \$64 million before abandoning their plant at Morris, Illinois.[41] President Ford's reappraisal of reprocessing in 1976 was partly made on economic grounds and in April 1977 his successor, President Carter, announced that the plant at Bradwell, South Carolina, would receive neither federal encouragement or funding for its completion as a reprocessing facility. In both the USA and UK, research and development costs were originally absorbed in military budgets for plutonium production but as civil programmes increased in size, nuclear power development became a charge on the overall civilian energy budget. In the UK the Department of Energy gives the Atomic Energy Authority a separate nuclear energy vote. In 1975/76 this vote accounted for over half of the Government's expenditure on research and development of all energy resources. As a result of Government subsidies, the low cost per kW often quoted by the CEGB in its reports for nuclear power stations must be viewed with caution.

The Central Policy Review Staff's proposal for export assistance seems to suggest that the Government should license PWRs to take advantage of their dominant position in the world market. At the beginning of 1977 the world market was worth \$20,000 million with 54 reactors, overall capacity 40,000 MW, under construction for export orders.[42] However, both the economics and the ethics of nuclear sales abroad have been seriously questioned. The dominance of PWRs has been achieved at uneconomic prices. By November 1976, for example, Westinghouse had lost \$1,000 million on its nuclear activities, not including its fuelling contracts. In the US reactor orders plummeted from 36 in 1973 to 27 in 1974 and 4 in 1975.[43] To prevent the collapse of the power manufacturing industry, export sales have been arranged through the US Export-Import Bank which has given loan guarantees, low interest charges and deferred payments to customers.[44] The ethics behind sales of this nature are disturbing. Apart from the possibility of nuclear proliferation, which will be discussed in more detail later, many of the buyers of PWRs do not require nuclear power. John Hill, Chairman of the UK Atomic Energy Authority, admits that it is not in the interests of the developed countries to push nuclear power into the underdeveloped countries at present.[45] He adds that most of these countries are more suited to the development of alternative, including renewable, energy resources. Indeed, the transfer of nuclear technology seems to run against recent attempts in the underdeveloped countries to move away from capital-intensive technology. Countries which accept nuclear technologies increase their reliance on industrial nations where there is a high concentration of the world's fuel cycle technologies and skilled technicians. There cannot be a more inappropriate technology for parts of the Third World with acute problems of over-population, under-employment and adequate alternative natural energy resources than the adoption of nuclear power.

Apart from the moral arguments, Hancock points out that no country should attempt to install a single power plant which would represent more than 15 per cent of the country's electricity grid capacity.[46] For example, only countries with 4,000 MW capacity already installed should consider buying a 600 MW reactor – Argentina, Brazil, Egypt, India, S. Korea, Mexico and Venezuela come into this category. In the light of the economic and ethical arguments against the sale of reactors in the Third World, should the Policy Review Staff's recommendation for export assistance be heeded?

Additional 'hidden costs' of nuclear plants are the costs of insurance and the de-commissioning of plants at the end of their working lives. The insurance issue gives nuclear power a distinct commercial advantage over its energy supply competitors, which have to provide third-party insurance out of their own funds. The nuclear industry is protected by national and international funds. In the early years of development it was felt that nuclear power was more of an unknown quantity than today. It would clearly have inhibited the development of nuclear power for peaceful purposes to have imposed on commercial operators and suppliers of components an unlimited liability which they would not be able to cover by insurance and which would, in so far as an operator was unable to meet his liability, confer no benefit on the victims.[47] The 'safety net' for nuclear operators came in the form of the Nuclear Installations Acts of 1959 and 1965. Initially, in 1959, nuclear operators had to take out insurance cover for £5 million per incident and, if a claim arose for more than this amount, the Government would pay up to £43 million. The latter figure was raised to £45 million in 1965 with support from international funds. Although the Department of Energy states that the limits are kept under review, operators continue to pay for cover of £5 million per incident, the same figure after nearly 20 years of operation. This is hardly in line with developments elsewhere in the insurance market. Insurance companies are quick to point out the inadequacies of under-insurance in the property market in times of high inflation.

Another facet of nuclear power which is frequently overlooked is the cost of de-commissioning reactors. Only 20 reactors in the western world have been closed down, most being small prototype reactors like the PFBR at Dounreay. By 2000, however, the Energy and Research Committee of the EEC predicts that 45 plants in the community will have reached the end of their useful lives and Guido Brunner, the Energy Commissioner, has proposed a £7 million, 5-year research programme into the problems of dismantling them.[48] For a large reactor, the cost and time involved in such a delicate operation is predicted to be between $23 and $64 million over a period of 48 to 108 years. This is equivalent, at 1975 prices, to between 4 and 13 per cent of the original plant cost.

If the 'hidden costs' of nuclear power are considered, nuclear power is not an attractive economic proposition in the 1970s. This has been appreciated by President Carter, who hoped that other countries would assess alternative energy sources 'if for no other reasons than because of economic consider-ations.'[49] In the UK the choice of electricity generating plant depends on fuel and capital costs, the anticipated length of time in operation and the yearly output. The main difference between conventional and nuclear stations is that fuel costs account for two-thirds of generating costs in fossil-fuel fired plant, but only one-fifth in nuclear plant.[50] Because of high capital costs to construct nuclear power stations and low running costs, they are costed to provide base-load electricity throughout their working lives. A fixed parameter for nuclear stations is a 75 per cent load factor; however, the average load factor for the Magnox stations from 1967 to 1972 was only 58 per cent.[51] This mediocre performance has improved since 1972 and most of the Magnox stations commissioned in the 1960s achieved consistently high load factors (80 to over 90 per cent) in the mid-1970s.

Unfortunately, the successors to the Magnox design and the last of the Magnox generation – Wylfa – have not come up to expectations. The Magnox stations were constructed during a period of low inflation when interest charges were low – an important economic consideration in the planning of nuclear stations. In the 1970s capital costs have escalated. Even the AGR programme, which began in a period of low inflation, has been affected by construction delays and hence higher costs. Henderson suggests that the technical problems which caused the delays and escalation in costs are peculiar to AGRs and the LWRs could have been built more quickly.[52] On the other hand, while a programme of PWRs may have saved capital costs, their reliability is questionable. In 1975 the US Nuclear Regulatory Commission announced that 42 commercial nuclear plants in operation in 1974 had an average load factor of 57.2 per cent. Moreover, the performance of plants became poorer with age, rather than better, because of premature ageing as a result of corrosion and fatigue. By their seventh year, PWRs were achieving a load factor below 40 per cent.[53] However, fossil fuel plants have also encountered problems: Drax A and Kingsnorth, two of the newest plants, recorded load factors of 59.5 and 54.3 respectively in 1976 owing to generator problems. The unreliability of nuclear plants presents a more acute problem because they are costed to operate on base-load throughout their lives. Wylfa, the largest and last of the Magnox programme commissioned in 1971, has been consistently beset with difficulties and its load factor of 23.1 in 1976 is fairly representative of its performance in the 1970s. Clearly, nuclear costs and thus the overall electricity cost structure will not be checked unless reliability is improved. To quote only one example, the Hunterston B shutdown in 1977 cost £16 million for repairs and the necessity to bring old plant into operation to replace it.[54] A rise in electricity prices to combat cost increases can only depress demand further – a situation the CEGB and the Government wish to avoid.

Nuclear power and the environment

Although fossil fuel plants are guilty of polluting the environment, nuclear stations have environmental problems which are unique in the generating of electricity. The main points of concern expressed in the Flowers Report and acknowledged by nuclear scientists are those of storing high level waste, the problem of nuclear proliferation and the safety of nuclear installations.

The main issue in the 1970s is undoubtedly that of waste storage. Flowers was confident that an acceptable solution to waste storage would be found but warned against a commitment to a large nuclear programme if this could not be proven beyond all reasonable doubt.[55] BNFL is constructing a £40 million glassification plant at Windscale to perfect a process introduced by Harwell scientists in the early 1960s.[56] This 'harvest' technology will vitrify high-level waste to facilitate its ultimate disposal. If this technology has been known for so long, why has it not been perfected before now? Contributory factors could be the lack (until recently) of environmentalist lobbying, the small amounts of waste involved (which did not create a major disposal problem compared with future contracts for foreign waste), or perhaps that the vitrification of waste is proving more difficult than is often claimed. It must be remembered that 99.9 per cent of the radioactivity of all nuclear wastes decays within 10 years[57] and it is only the high-active wastes such as plutonium with a half life of 24,400 years that present a major long-term disposal problem.

Research is increasingly being focused on reception areas for this waste. The

National Radiological Protection Board (NRPB), Britain's watchdog agency for public exposure to radiation, feels that all of the world's nuclear wastes by 2000 can be stored conveniently in a geologically quiet region of the ocean floor, deep enough to avoid interaction with trawling activities and removed from undersea cables.[58] Nevertheless, it is more likely that the waste will be buried underground in suitable geological areas. Chapman, Gray and Mather feel that this method is more practical with the vitrified waste being stored 300 to 1000 metres under the earth's surface serviced by inclined tunnels and shafts.[59] To prove that this method is foolproof, they point to the natural example of a uranium mine at Gabon where a natural fission took place in Pre-Cambrian times and containment has occurred throughout 1800 million years of geological conditions. In the UK the geological structure is sufficiently diverse to provide many areas as potential hosts for nuclear waste. The self-sealing properties of clay and salt rocks show possibilities although the salt domes studied in West Germany and the USA are much thicker and are only present in the same form in the UK under the North Sea. It is more likely that hard crystalline, igneous or metamorphic rocks, such as are found in much of Highland Britain, will be the focus of most research and development because the rocks are dry, stable and relatively inert. Understandably, people who live near areas earmarked for nuclear waste storage are objecting to UKAEA proposals for exploratory drilling.[60]

In order to put the controversial issue of ionising radiation in perspective, it must be noted that nuclear wastes only account for 0.01 per cent of the annual radiation dose in the UK.[61] Natural sources, including radioactivity in the human body, account for 84 per cent of the annual dose whilst artificial sources occur mainly in the form of medical irradiation (13.5 per cent) and fallout (2.1 per cent). Overall, in the UK, the population receives a dose of between 80 and 150 millirads of ionising radiation per year.[62] Susceptibility to radiation exposure depends more on one's occupation – airline crews receive more cosmic rays than average, miners more radiation from rocks than average – and place of residence – some areas will have higher natural radioactivity because of altitude and the types of building material used. However, a complacent attitude towards 'controlled' emissions of radioactive gases and wastes into the environment from nuclear activities is undesirable. A government committed to a programme of nuclear expansion, especially in its reprocessing activities, requires to monitor radioactive emissions strictly. Throughout the fuel cycle, the production of significant amounts of gaseous, liquid and solid wastes are subject to constant monitoring by the Department of the Environment and the Ministry of Agriculture, Fisheries and Food.

Radioactive wastes from reactors are released into the environment through the activation of argon, which occurs naturally in the air, and the mixing of fission products into the coolant. Other releases are related to the short-term storage of irradiated fuel. The water in the storage ponds is mixed with cooling water before being discharged into the environment. Radiation doses are much higher around Trawsfynydd, situated on a freshwater lake, than other reactor sites where coastal waters make the dispersal of wastes easier.

Most environmental problems are associated with emissions from reprocessing plants. The time lag between irradiation in the reactor and reprocessing results in the decay of most of the fissile products. One gas – krypton 85, with a half-life of 10.76 years – is an exception and Mr Justice Parker recommends that this gas should no longer be released to the atmosphere but retained. Other products with short half-lives are released into the Irish Sea from Windscale at a rate of 500,000 litres a day.[63] An alarming aspect of these discharges is the steep rise in caesium 137 releases (30 times from 1960 to 1970)[64] and other beta-ray emitting nuclides to reach 82 per cent of the authorised limit.[65]

Much concern was expressed at the public inquiry about the thoroughness of monitoring procedures and doubts were expressed over the legitimacy of 'the acceptable dose'. Cumbria as a region, and Windscale workers in particular, could be subjected to unacceptable doses of radioactivity. The inquiry highlighted some of the inadequacies of radiological protection. Tests were carried out on locally caught fish, potatoes from the Isle of Man and silt in the Ravenglass estuary. Lakes Thirlmere, Haweswater and Wet Sleddale were checked for tritium after fears were expressed over the possible contamination of Manchester's water supply. Whether such short-term tests are meaningful is open to dispute but regardless of the results, why did it require a public inquiry to initiate action of this nature?

Brian Wynne not only questions monitoring procedures but levels criticism at the bodies which are responsible for establishing radiation dose limits.[66] For example, the guidelines of the International Commission on Radiological Protection (ICRP) have been abandoned in the US, where an independent agency has been created to measure scientifically the threshold dose which can give zero risk of harm. From this scientific base, the Environment Protection Agency can produce a cost-benefit analysis on the likely costs which would be incurred to avoid additional exposures. As a result, the American standards are 100,000 times stricter than their British equivalent, where ICRP judgements are the accepted norm. Wynne argues that the NRPB, the authoritative body at national level, is obviously not unbiased as most of its members are ex-AEA scientists. Mr Justice Parker has recommended that an independent body with interests in protecting the environment should be created to advise the Government on fixing radiological protection standards.

Although preliminary studies by the NRPB indicate that Windscale workers are at no greater risk from cancers than the population at large,[67] the data base is incomplete and the method of analysis has been severely criticised.[68] The controversy over this issue will not be resolved for about a decade. By that time statistics on the missing information – workers who have left the industry – will provide a base upon which to make a meaningful hypothesis. More meaningful statistical trends will be discerned in the 1980s with the introduction of an accurate radiation register for all UK workers.

In the US a radiation register was established in the 1960s and, with the aid of social security documentation, follow-up studies have been undertaken of workers who have left the industry. Dr Alice Stewart, a leading epidemiologist, has studied records, dating back to the 1940s, of the Hanford works in Washington, the US equivalent of Windscale. If the results of this survey are applied to Windscale, they seem to suggest that the risks from low-level radiation may be much greater – up to 20 times greater – than current estimates indicate.[69] A recent study of naval shipyard workers at Portsmouth, New Hampshire, endorsed the concern expressed by Dr Stewart.[70] Workers repairing nuclear submarines were much more susceptible to death from cancer than non-nuclear workers. (34.4 per cent compared with 21.7 per cent). These figures for the total sample underemphasise the increased susceptibility of particular age-groups. For example, in the 60 to 69 age group, 60 per cent of nuclear workers died of cancer compared with 20 per cent of non-nuclear workers. Clearly, the correlation between the level of radiation exposures and the length of time in the industry is significant and raises considerable doubts about the long-term safety of Windscale workers.

Nuclear proliferation

As nuclear plants were originally conceived to manufacture nuclear weapons and not to produce electricity, an expansion of the UK's nuclear programme is often deemed to be a step towards the 'plutonium economy'. The Fox Commission studying the environmental implications of further uranium mining in Australia points out that 'the nuclear power industry is unintentionally contributing to an increased risk of nuclear war'.[71] Uranium exporting countries and others who have developed 'sensitive technologies' – uranium enrichment, reprocessing and heavy water refinement – have become increasingly concerned about the misuse of nuclear technologies. In North America, proliferation has become a major political issue following India's 'peaceful' explosion developed from imported Canadian power technology. The US policy towards reprocessing to minimise proliferation can have political repercussions in other countries. Japan and West Germany are users of US-enriched fuel and they require American authorisation to reprocess this fuel. In view of the Japanese contract with the UK, it is not guaranteed that the Japanese can send its irradiated fuel to the UK. The Canadians, clearly worried after their Indian experience, suspended shipments of uranium to the EEC for a year whilst safeguards were renegotiated.[72]

Although the International Atomic Energy Agency (IAEA) has tightened its safeguards to ensure that nuclear technology is used and exported only for peaceful purposes, a country committed to nuclear re-armament can easily flout the rules. The IAEA inspection systems can detect whether nuclear materials have been diverted for military purposes but they cannot prevent their misuse.[73] The present international safeguards through the IAEA and Non-Proliferation Treaty (NPT) are inadequate. Matters are not helped by the irresponsibility of countries such as West Germany and France, which have made nuclear deals with Brazil and Pakistan respectively, neither of which have signed the NPT. The German action has been discredited because it involves a 'package' deal – the export of not only reactors but enrichment and reprocessing technology. Considering the dubious financial benefits derived from reprocessing, it is more likely that imported plant will be used to extract plutonium for a nuclear weapons programme.

Proliferation may also result from the 'dumping' of reactors on world markets. Professor Wohlstetter claims that reactor grade fuel can be used to make a crude bomb.[74] The efforts of the US and other industrialised countries to sell reactors worldwide could be contributing to proliferation. As western developed countries re-assess the nuclear situation, the eastern and developing states – for example, the USSR, and S. Korea – are building up their nuclear programmes.[75] Public opinion in Canada has led to the suspension of the sale of Candu reactors to Argentina. Nevertheless in 1977 Brazil agreed to cooperate with Argentina on nuclear matters. Is this a 'backdoor' method by which

Argentina can obtain West German technology? The IAEA forbids the re-export of nuclear technology but as both countries have not signed the NPT, they will probably disregard IAEA guidelines.

It is possible that the construction of fewer, larger plants, such as THORP, can minimise the world production of plutonium. However, strict security arrangements would be necessary to guard vast stockpiles of plutonium, and additional safeguards would be imperative in the transportation of fuel to and from reprocessing plants. Two issues would arise from such a situation. Firstly, the increased likelihood of theft or sabotage of the fuel en route and, secondly, the erosion of civil liberties which would result from an increase in security arrangements. The Windscale planning permission, for instance, involves the vetting of employees and restrictions of trade union rights.[76]

Although no figures are available for 'materials unaccounted for' because it is classified information, uranium thefts have occurred in the past. The most notorious case is that of the disappearance of a shipment of 200 tonnes of oxidised natural uranium which left Antwerp on 16 November 1968 and never arrived at its intended destination in Italy. Instead, the uranium was apparently diverted to the Dimona reactor in Israel, a country desperate for fuel after the French discontinued supplies during the 1967 Arab-Israeli war.[77]

Mr Justice Parker suggests that the problem of sabotage of plutonium can be resolved by returning it in the form of irradiated fuel rods which can be used only as reactor fuel. This recommendation is highly impractical, the technology of 'spiking' is expensive and commercially complex as the fuel would have to be designed for individual reactor specification; for example, the fuel returned to Japan would have to be specially fabricated to suit Japanese reactors.[78]

The development of nuclear technologies seems to have led the world along a less peaceful path. As Robin Cook MP points out, the talk of the 1970s is not about nuclear disarmament but nuclear 'arms limitation'; if a so-called responsible, stable, democratic country such as Britain could carry out two nuclear tests in 1974 and 1976 how can emerging nations such as Brazil and Iran be expected not to follow our example?[79]

Nuclear safety

Although much of the discussion on nuclear safety revolves around the 'controlled' releases of radioactive wastes into the environment, the possibility of an unexpected accident cannot be dismissed. In Britain the administrative machinery exists to plan and design nuclear establishments within the framework of stringent safeguards. After the Windscale fire of 1957 when both production reactors were prematurely closed down, the Nuclear Installations Act was introduced. The Nuclear Installations Inspectorate (NII) was created and given the responsibility to license plant and to ensure that high safety standards were maintained during operation. By 1974 the NII had come under the umbrella of the Health and Safety at Work Act with the Health and Safety Executive assuming responsibility for health and safety in the nuclear industry.

Statistically, the nuclear industry's record for safety is better than those of most other industries. In 1976, for example, the number of fatal accidents per thousand employees in the fuel industries was:[80]

Coal mining	0.19

Offshore Oil and Gas	1.5
Electricity	0.09
Gas	0.10
Nuclear	0.00

It was shown in Chapter 3 that the oil industry often plans for safety *last*, and so the inadequacies of one industry on safety standards should not be used as a yardstick to measure the nuclear industry, which has unique safety problems. Shorthouse,[81] in his evidence at Windscale, argued that the proposed reprocessing plant was not an ordinary chemical plant. If its engineering system failed, the result could not be compared with accidents such as aircraft disasters or the Flixborough explosion, which only affected lives and property for the duration of the event. A nuclear accident could have repercussions on future generations and must therefore be viewed accordingly in any comparative analysis.

It would be foolish to dismiss the possibility of a major nuclear accident. Dunster admits that various kinds of accidents, graded in severity, will probably occur in the long term.[82] Sir Kelvin Spencer, former Chief Scientist to the Government, is more blunt on the issue, claiming that the question was not whether there will be an accident, but when.[83]

It is the job of the NII to ensure that the quality of the design of reactors and the engineering standards are sufficiently high to minimise the risk factor. Although Flowers was sufficiently confident of reactor safety to recommend their siting in urban areas to utilise waste heat, much depends on the choice of reactor.[84] The Magnox and AGR designs are considered relatively safe compared with the PWR, which has been the subject of extensive debate. In the US public opposition to licensing PWRs in the early 1970s stimulated the publication of a series of safety studies sponsored by the Atomic Energy Commission. The most quoted was the Rasmussan Report, 'An Assessment of Accident Risks in US Commercial Nuclear Power Plants', of 1974, which Patterson cynically dubs a child's guide to nuclear safety.[85] Rasmussan calculated the risks and chances of different kinds of accidents occurring. The worst possible PWR accident, defined as a category 1 accident, would be a break of the coolant pipe or rupture of the pressure vessel; if the safeguards failed to work, the reactor core would melt through the vessel and the floor of the building. The odds on this happening are placed at once every million reactor years: however, as Dunster points out, this is a small risk at the present scale of nuclear power development – once every 3,000 years – but, with the expansion of nuclear programmes the risk probability will increase to once every 100 years by the end of the century.[86]

The PWR controversy rapidly spread across the Atlantic when it was revealed that the CEGB intended to base its new generation of reactors on this design. Hostility to this decision built up, especially when Sir Alan Cottrell, who had just retired from his post of Chief Scientific Advisor to the Government, sided with the objectors. Sir Alan, a leading metallurgist, wrote a strong letter to the *Financial Times* in June 1974 which concluded 'when the consequences to the general public of such a failure are as uniquely grave as in the reactor case, it would be wise to choose a system less critically dependent upon human perfection than the steel pressurised water reactor.'[87] Cottrell's main doubts concern the ability of the steel in the pressure vessel to withstand changes in temperature and pressure throughout its working life. In the wake of this criticism, the UKAEA under Dr W. Marshall undertook a two-year study of the safety of PWRs.[88] The report found no basic difference in technology and therefore safety between the PWR and the SGHWR and saw no reason why PWRs could not be licensed for construction in Britain. Cottrell is still far from convinced that PWRs should be a part of Britain's nuclear plans for the immediate future. His main argument is that alternative reactors are available which could be located close to urban areas, unlike PWRs, which require a higher standard of inspection than other designs.

The reactor of the future, the FBR, is a more sophisticated engineering design but doubts have been expressed about its safety.[89] The NII, when questioned by Friends of the Earth on FBR safety, replied that an explosive reaction could occur if the core was disrupted and the molten fuel reacted with liquid sodium, the coolant. The Inspectorate calculated that the 'worst credible accident' from a FBR in an urban location would result in several thousand deaths within a few weeks of the accident in an area extending six miles downwind of the reactor.[90] Clearly, the population exposed to radiation would extend for a few hundred kilometres downwind, although a lethal dose would not necessarily be inhaled.

The likelihood of an accident at any point in the fuel cycle is remote, and the 'worst credible accident' is most likely to come from a reprocessing plant. In Britain many 'incidents' have been recorded at Windscale but only two major accidents have occurred. The fire of 1957 caused radioactive contamination to the surrounding area which necessitated the throwing away of milk; in 1973, an unexpected chemical reaction took place due to a build-up of ruthenium 106, and 35 personnel were exposed to skin and lung contamination from this beta-emitting radioisotope. The 'head end' plant, which had been converted to reprocess oxide fuel in 1969, had only processed 100 tonnes. Even Mr Hill, Chairman of the AEA, admitted that 'we underestimated the problem for the reason that we used the wrong unit of throughput for our extrapolation'.[91] He conceded that Magnox fuel is much easier to handle than the oxide fuel from the new generation of reactors.

It is almost certain that major accidents have occurred in the USSR which can be attributed to some kind of reprocessing or waste disposal plants. Dr Zhores Medvedev, the dissident Soviet scientist, revealed details of an explosion in the South Urals region in the winter of 1957/58.[92] He claimed that thousands of people were affected, hundreds were killed and a vast area was closed to the public. From his research, he deduced that an atomic explosion or a reactor accident was not responsible for the devastation but an accident related to nuclear wastes. By documenting Soviet research throughout the 1960s, he pieced the jigsaw together to show that scientists were studying the genetic effects of strontium 90 and caesium 137 in an unnamed site, which had similar characteristics to the South Ural area. His case was endorsed by Lev Tumerman, formerly head of Biophysics at Moscow, who had visited the area between Chelyabinsk and Sverdlovsk in 1960 and found miles of barren, unusable land.

More recently, in October 1976, an 'earthquake' was reported at the Estonian naval base at Paldiski. The 4.5 Richter scale 'earthquake' has been interpreted as a nuclear explosion as it is believed that nuclear silos are stored at Paldiski.[93] This claim could have some justification in that Finnish scientists have reported increased radioactivity levels in the Gulf of Finland.

In West Germany, the Institute of Reactor Safety at Cologne has considered the worst credible accident that could occur in a waste reprocessing plant.[94] It calculated that technical failures could lead to radioactive fallout killing everyone within 100 kilometres downwind of the plant – if the plant was located in North Germany and a north wind was blowing at the time. These figures have been disputed by the Political Ecology Research Group[95] but, regardless of the uncertainties of the data, has the NRPB or NII accurately estimated the worst accidents that could happen at Windscale in order to minimise *all* potential risks to the nearby population?

Conclusion

In an attempt to answer the question posed at the beginning of the chapter, from the evidence produced it seems likely that nuclear power will be the cornerstone of British energy policy in the twenty-first century. Whether it will be our salvation or damnation is open to question. Britain will need nuclear energy and undoubtedly its use can preserve living standards at close to present levels, but the rate of development must be scrutinised. The public electricity supply is presently suffering from overcapacity and even if demand begins to pick up, it will be unnecessary to order any new power stations until the late 1980s. If the main reason for immediate nuclear expansion is to keep the power manufacturing industries alive, other methods of support than the construction of new plant are available to the Government. It has been suggested that power manufacturing industries could channel their efforts into improving reliability of modern plant and increasing the thermal efficiency of smaller, old stations.

Will the nuclear future be safe? While countries such as Denmark, less well endowed with fossil fuel resources than the UK, can afford to discount the nuclear option, the British Government is contemplating a programme of nuclear expansion with unproven and perhaps unsafe technology. PWRs are not reliable, and their safety is open to doubt, yet strong pressures exist within the nuclear industry to adopt this reactor in future planning.

The Windscale inquiry did nothing to allay fears on the problems of reprocessing. Mr Justice Parker feels that he cannot make any judgement on the safety of THORP; that he cannot assess the risks of nuclear proliferation; that he can see no solution to the problem of the erosion of civil liberties and that he cannot reach a conclusion on the morality of leaving it to future generations to decide whether to monitor or dispose of wastes.[96] And yet, planning permission was recommended. Can BNFL justify the construction of the largest reprocessing plant in the world after its limited, and unfortunate, experience in handling oxide fuel? The dubious economic benefits seem to be the only justification. Small-scale plants are uneconomic, and BNFL needs to build a plant of sufficient size to reprocess foreign fuel in addition to domestic fuel to derive scale economics. Even then, the economics of the plant are questionable.

North Sea oil and gas have given Britain time to develop 'safe' technologies for the future. Nuclear power could be safe, given time, money and resources for research. Perhaps the initiative for perfecting nuclear technologies should be left to countries with more acute energy problems and greater financial resources for research and development than the UK.

Notes and references

1. Royal Commission on Environmental Pollution, Sixth Report (1976) *Nuclear Power and the Environment* (Flower's Report), HMSO.
2. Select Committee on Science and Technology, First Report (1976) *The SGHWR Programme*, HMSO, para. 43.
3. Department of Energy (1977a) *United Kingdom and Community Energy Policy*, Department of Energy, p. 6.
4. Town and Country Planning Association. *Annual Report*, 1978, pp. 13, 14.
5. J. Hill (1975) 'The energy situation and the role of nuclear power', *Atom*, No. 219, January 1975, p. 6.
6. Energy Commission, *Working Document on Energy Policy, Paper No. 1*, pp. 35, 36.
7. The Electricity Council (1973) *Electricity Supply in Great Britain: A Chronology*, Electricity Council, p. 61.
8. W. C. Patterson (1976) *Nuclear Power*, Pelican, p. 140.
9. See M. Gowing (1974) *Independence and Deterrence: Britain and Atomic Energy 1945–1952*, vol. 2, Macmillan.
10. The US and USSR would possibly dispute this claim because they had small power reactors in operation before this date.
11. P. D. Henderson (1977) 'Two British errors: their probable size and some possible lessons', *Oxford Economic Papers* vol. **29**(2) July 1977, pp. 159–205.
12. *Financial Times*, 23 February 1976.
13. *Financial Times*, 8 March 1977.
14. The SGHWR design incorporates features of the Canadian Candu reactor which uses heavy water as a moderator and the BWR which uses enriched uranium fuel.
15. UK Atomic Energy Authority (1976) *An Assessment of the Integrity of PWR Pressure Vessels*, UK AEA.
16. W. C. Patterson (1978) 'Nuclear AGRo breaks out again', *New Scientist*, 2 February 1978, p. 286.
17. Central Policy Review Staff (1976) *The Future of the United Kingdom Power Plant Manufacturing Industry*, HMSO, p. 95.
18. See Patterson (1978) op. cit., pp. 75–77.
19. C. Conroy (1978) *What Choice Windscale?* Friends of the Earth/Conservation Society, p. 23.
20. Flowers Report, op. cit., para 533.
21. *Guardian* (1977), *Windscale: A Summary of the Evidence and the Argument*, a *Guardian* pamphlet, pp. 17, 18.
22. Ibid, p. 16.
23. Mr Justice Parker (1978) *The Windscale Inquiry*, HMSO pp. 28–31, 37, 38.
24. Mr Justice Russell Fox (1976) First Report, *The Ranger Uranium Environmental Inquiry*, Australian Government Publishing Service, October 1976.
25. Nuclear Energy Policy Study Group (1977) *Nuclear Power: Issues and Choices*, Ballinger.
26. *Planning Bulletin*, 24 February 1978.
27. *Financial Times*, 21 February 1977.
28. *Financial Times*, 4 February 1976; and A. Gibson, (1977) 'The Jet Project' *Atom*, No. 254, December 1977, p. 337.
29. KAD Inglis (ed.) (1973) *Energy: From Surplus to Scarcity?* Ch. 8 by D. C. Leslie, Applied Science Publishers Ltd, pp. 129 and 130.
30. Department of Energy (1977b) *Digest of UK Energy Statistics 1977*, HMSO, Tables 66 and 68.
31. R. Vale (1978) 'All-electric lies', *Undercurrents* No. 26, February-March 1978, pp. 12, 13.
32. Department of Energy (1977c) *Compare Your Home Heating Costs*, Department of Energy, pp. 7–18.
33. Central Policy Review Staff (1976) op. cit.
34. Patterson (1978) op. cit., p. 287.
35. *Financial Times*, 29 April 1977.
36. G. Manners, letter to *The Times*, 25 January 1978.
37. G. Manners (1977) 'Alternative strategies for the National Coal Board', a paper presented at a conference entitled *Energy and the Environment* Regional Studies Association, May 1977.
38. W. C. Patterson quoted in the *Yorkshire Post,* 21 June 1977.
39. H. A. Bethe (1976) 'The necessity of fission power', *Scientific American*, **234**, January 1976, p. 30.
40. *Financial Times*, 5 March 1975.

41. J. Bugler (1976) 'Windscale's fuel for controversy', *New Statesman*, 26 November 1976, pp. 740, 741.

42. F. Barnaby (1977) 'The politics of reprocessing', *New Scientist*, 6 April 1977, p. 18.

43. G. Hancock (1978) 'The hatching of the nuclear bird', *New Internationalist*, No. 61, March 1978, p. 5.

44. Ibid. and Patterson (1978) op. cit.

45. J. Hill (1977) in 'Nuclear Power and the Energy Future', *Atom*, No. 254, December 1977, p. 361.

46. Hancock (1978) op. cit., p. 6.

47. Department of Energy (July 1976) *Nuclear Activities in the UK: Commentary on Some Points of Interest*, Department of Energy, p. 5.

48. *The Economist*, 6 May 1978, p. 64.

49. Quoted in Conroy (1978) op. cit., p. 27.

50. G. L. Reid, K. Allen and D. J. Harris (1973) *The Nationalised Fuel Industries*, Heinemann, p. 197.

51. C. Sweet (1978) 'Nuclear power costs in the UK', *Energy Policy*, June 1978, pp. 104–7. A plant load factor is the average hourly quantity of electricity supplied during the year expressed as a percentage of average output capacity during the year. In the US this is often referred to as the 'capacity factor'.

52. Henderson (1977) op. cit.

53. Patterson (1976) op. cit., pp. 225, 226.

54. *Yorkshire Post*, 15 June 1978.

55. Flowers Report, op. cit., paras 181 and 338.

56. F. Feates and N. Keen (1978) 'Researching radioactive waste disposal', *New Scientist*, 16 February 1978, p. 426.

57. P. Beckmann (1977) (L. G. Brookes reviews) 'The health hazards of not going nuclear', in *Atom*, No. 244, February 1978, p. 20.

58. National Radiological Protection Board (1977) *Assessment of the Radiological Protection Aspects of Disposal of High Level Waste on the Ocean Floor*, NRPB, R48, HMSO.

59. N. Chapman, D. Gray and J. Mather (1978) 'Nuclear waste disposal: the geological aspects', *New Scientist*, 27 April 1977, p. 226.

60. *Planning*, No. 284, 8 September 1978.

61. Department of Energy (July 1976) op. cit., p. 26.

62. H. J. Dunster (1975) 'The Atom and the environment', Ch. 5 in J. Lenihan and W. W. Fletcher (ed.) *Energy Resources and the Environment*, Blackie, p. 131.

63. Patterson (1976) op. cit., p. 107.

64. P. Bunyard (1976) 'Gearing up to the plutonium economy', *Ecologist*, **6**, p. 348.

65. *Guardian pamphlet* op. cit., p. 83.

66. B. Wynne (1978) 'Politics of nuclear safety', *New Scientist* 26 January 1978, pp. 208–211.

67. G. W. Dolphin (1977) *A Comparison of the Observed and Expected Cancers of the Haematopoietic and Lymphatic Systems among Workers at Windscale*, First Report, NRPB, R54.

68. By Sir R. Dell, a member of the Royal Commission on Environmental Pollution, see the *Guardian* (1977), op. cit., pp. 94, 95.

69. Ibid, p. 72.

70. L. Torrey (1978) 'Radiation haunts shipyard workers', *New Scientist*, 16 March 1978, pp. 726, 727.

71. Quoted in G. Boyle (1977) 'Do we need the half life of the plutonium economy?' *Undercurrents*, No. 22, June/July 1977, p. 7.

72. *The Economist*, 24 December 1977, p. 43.

73. An IAEA report leaked to Dutch Friends of the Earth was reported in the *Observer*, 25 June 1978. It admitted that inspections were difficult to carry out and that safeguards could be broken.

74. Conroy (1978) op. cit., pp. 62, 63.

75. *The Economist*, 'Nuclear man at bay', 19 March 1977, pp. 12, 13.

76. *Guardian pamphlet*, op. cit., pp. 93, 94; and Boyle (1977) op. cit., p. 7.

77. *Sunday Times*, 8 May 1977. A *Panorama* programme also featured this hijack on 26 June 1978.

78. Conroy (1978) op. cit., p. 64.

79. R. Cook (1976) 'Against the British nuclear capacity', *New Statesman*, 1 October 1976, p. 440.

80. *Atom*, No. 251, September 1977, p. 195.

81. *Guardian pamphlet*, op. cit., p. 82.

82. Dunster (1975) op. cit., p. 150.

83. Sunningdale Seminar, 13 and 14 May, reported in *Atom* No. 251, September 1977, p. 195.

84. Flowers Report, op. cit., paras 176 and 295.
85. Patterson (1976) op. cit., p. 202.
86. Dunster (1975) op. cit., pp. 151, 152.
87. In D. Fishlock 'A reactor controversy', *Financial Times*, 8 October 1976.
88. UKAEA (1976) *An Assessment of the Integrity of PWR Pressure Vessels*, (Marshall Report), UKAEA.
89. Patterson (1976) op. cit., pp. 181, 168, 177.
90. *Guardian*, 14 January 1977.
91. J. Hill (1976) 'Radioactive waste management in the United Kingdom', *Atom* No. 240, October 1976, p. 250.
92. Z. Medvedev 'Two decades of dissidence' *New Scientist*, vol. 72, p. 264; and 'Facts behind the Soviet nuclear disaster', *New Scientist*, vol. 74, pp. 761–4.
93. South Yorkshire Nuclear Action Group, *Nuclear Incidents and Events, October 1976–June 1977*, p. 4.
94. Institute of Reactor Safety, *Report No. 290*, IRS 1977.
95. Political Ecology Research Group Oxford (1977) Report No. 2: *Use and Abuse of Information in the Nuclear Power Debate: a German Case Study and its Implications for Europe*, PERG.
96. Mr Justice Parker (1978) op. cit., paras 11.7, 6.33, 7.12 and 13.6.

Chapter 5

Renewable energy resources

There are two basic renewable sources of energy available to us:

– Radiation from the sun, which can appear as ambient heat, as energy transferred to the wind, to the ocean waves, to water at high levels deposited from clouds or as energy stored in plant materials.
– The earth's rotation which through complex interactions between the sun, earth and moon can appear as tidal energy. [1]

The United Kingdom is geographically well-suited to the development of renewable energy resources. The Severn Estuary and the Outer Hebrides offer two of the best sites in the world for electricity generation by tidal and wave power respectively. The notion that this is 'free' energy should be placed in perspective by the realisation that numerous technological problems must be overcome in projects for the capture of renewable energy resources, in addition to the prerequisite of large-scale capital investment for their development. It might be argued that fossil fuels are also 'free' energy resources because they are *stored* solar energy;[2] the relative costs of development justify their exploitation at the expense of alternative forms of power.

The concept of 'alternative' energy is derived from the alternative technology movement, which questions the scale of existing technologies. Thus some large-scale renewable energy projects, such as the schemes to develop 600 miles of giant wave power generators along the coast and to erect thousands of 70-metre windmills are not strictly 'alternative' proposals. The 'alternative' approach is typified by the National Centre for Alternative Technology at Machynlleth, Powys, Wales, where low technology in the form of windmills, solar cells, a water wheel and solar panels are in use or exhibited. The mode of life at the centre is being practised in other rural communities which seek a move from centralised urban society to a more self-sufficient rural way of life. The centre could be a prototype for further ventures into the utilisation of small-scale renewable energy technologies. Nevertheless, in an overpopulated island such as Britain the necessary social and political changes associated with alternative technology required for the creation of rural communities as advocated by Dickson,[3] Boyle and Harper,[4] seem unlikely to materialise.

Regardless of the scale of technology, the various proponents of renewable energy are united in claiming that this energy is clean, safe and inexhaustible, compared with controversial nuclear power technologies. The Select Committee on Science and Technology felt that Government expenditure on renewable energy sources was inadequate; in 1976 £3.7 million was spent on research and development of renewable resources *plus* energy conservation, compared with £146.3 million spent on nuclear research and development.[5] The Committee did add, however, that these figures give no indication of progress, and many engineers would argue that until appropriate technological breakthroughs are achieved the Government is justified in its present allocation of finance for renewable energy research and development.[6] This chapter reviews the development of various renewable energy schemes, assessing their potential contribution to meeting UK energy demands by the turn of the century.

Hydro-electric Power

Hydro-electric Power (HEP) is a well-established renewable energy resource which makes a valuable contribution to electricity supplies in the more remote areas of Britain. Although electricity supplied by HEP has increased by a factor of 10 in the post-war period, consumption–especially in the 1960s–increased to such an extent that in 1976 HEP supplied only 2 per cent of UK electricity requirements.

The average generating capacity of the 71 conventional HEP stations is less than 20 MW; the largest of the 53 stations in northern Scotland, situated at Loch Sloy, has a generating capacity of 130 MW (see Fig. 4.1). The three pumped storage schemes currently in operation are much larger: Cruachan (400 MW), Ffestiniog (360 MW) and Foyers (300 MW). The principle behind these schemes is the self-utilisation of generated power. During off-peak supply periods water is pumped from a lower to an upper reservoir where it is stored until more generation is required. This is the only large-scale method of electrical bulk storage in use. The Dinorwic scheme in North Wales (at 1,840 MW, one of the largest in Europe) will be able to replace the failure of any turbo-generator in England and Wales, thereby giving the CEGB greater flexibility and security in controlling the national electricity grid.

Most potential sites in the UK for HEP grid-connected schemes are already being utilised, and an increase in demand for electricity will result in a proportional decline in the importance of HEP as a producer of power. The Centre for Alternative Technology claims that Britain could double its output of HEP by utilising small water-power sites.[7] Theoretically this is feasible, as a range of water turbines could be used in numerous sites, but the high cost of the turbine relative to the amount of delivered power would discourage its development, with the exception of a few isolated sites which may have a legacy of disused water wheels.

Solar energy

Various reports on the potential exploitation of solar energy have resulted in a marked change of attitude by the Department of Energy towards the role which solar power could play in a long-term energy strategy. The Department was criticised by the Select Committee on Science and Technology for complacency regarding the development of renewable sources of energy and the Committee cited the example of the Department's announcement of a solar energy programme in February 1977, 19 months after the completion of the research.[8] Nevertheless, all of the reports indicate the same technological and economic problems which have to be overcome before solar power can make a major impact on the energy market. The main differences concern the amount of money which should be allocated to a research and development programme, and the possible contribution solar energy could make to total energy supplies in the twenty-first century. The Department of Energy originally predicted that by the year 2000 the share of total energy demand supplied by solar energy would be less than 1 per cent. This has been upgraded to 2 per cent, with a possible tenfold increase in the longer term.[9] This optimistic forecast more than equals the UK section of the International Solar Energy Society's (ISES) forecast of 10 per cent of demand by 2020.[10] On comparing research and development budgets in other countries, ISES and the Select Committee felt that Britain should be spending more on solar energy research programmes. ISES proposed that Government funding should begin immediately, and should build up gradually to £10 million per year by 1980.[11] The Government has gone some way towards meeting this suggestion by making £3.6 million available to industry from 1977 to 1981 to assist the development and manufacture of domestic solar heating systems.

The Government, with some justification, is spending less on research and development than other countries because of the unfavourability of conditions in the UK compared with Australia, USA and Japan. Although total solar input per annum is barely half that of the USA or Australia, it is equivalent to 80 times present primary demand, but the variability of input presents storage problems. Much of the UK's solar input occurs during long summer days: unfortunately demand for energy is greatest in winter when insolation levels are low. It is not surprising that countries with small seasonal fluctuations in solar input and high mean insolation levels spend correspondingly more money than Britain on research and development. (Table 5.1).

Table 5.1 Insolation in various countries (J per m² per day x 10⁶) (*Source:* Department of Energy, *Energy Paper No. 16*, p. 21)

	Midsummer	Midwinter	Ratio	Annual mean
UK	18	1.7	10.0	8.9
Central USA	26	11.0	2.4	19.0
S. France	24	5.0	4.8	15.0
Israel	31	11.0	3.0	22.0
Australia	23	13.0	1.8	20.0
Japan	17	7.0	2.4	13.0
India	26	14.0	1.9	20.0

An added disadvantage for solar development in the UK is the high population density and consequently shortage of land for solar collection. ISES claims that at 30 per cent efficiency of conversion, 2.6 per cent of the UK's land area would be sufficient to meet total 1973 energy demand.[12] Which land? Undoubtedly a quantity of this land would be located in 'marginal' areas where land use competiton has already produced planning and environmental problems.[13]

Solar water-heating systems

The area of solar energy research which offers most promise for the future is the development of solar water heating systems. Already a great deal of research has been carried out, but existing commercial systems are too expensive and unreliable to secure large market penetration.

The solar collector consists of a black absorber panel in a well-insulated glass frame with water circulating throughout the system. Of the energy received, 7 per cent is reflected from the collector, but most of the incoming radiation (78 per cent) passes through to the panel, whilst the remaining 15 per cent is absorbed by the glass.[14] As the glass becomes warm, energy is re-radiated to both the inside and outside of the collector. The trapped heat energy is subsequently drawn off into a well-insulated storage tank. The efficiency of the system is dependent on the standard of insulation, the workmanship in the construction of the collector, the possible use of double glazing to reduce convection losses, the type of panel used, the rate of use of heated water and the distance this water has to flow to the storage tank. The amount of energy captured depends on the installation of the collector. Ideally, it should be positioned on a south-facing roof, at the best angle for collection of the incoming solar radiation. The optimal position varies from building to building according to its situation and its locational relationship to other structures. Vale and Vale assume that 40 per cent efficiency can be achieved and calculate that a 15 m² collector can gather sufficient energy to supply hot water for two people throughout the summer.[15] The figures in Table 5.2 are based on Kew records which indicate the range of insolation received from summer to winter months.

Table 5.2 Total energy collected (kWh) (*Source:* B. and R. Vale: *Autonomous House,* p. 61, Thames and Hudson 1975)

Jan	Feb	Mar	Apr	May	Jun
56.2	131.9	328.4	614.3	883.3	1036.4

Jul	Aug	Sept	Oct	Nov	Dec
758.0	729.4	413.6	223.8	78.4	42.4

This illustrates the disadvantages of the British climate in the development of solar systems. An area of potential market penetration for solar collectors in the UK is in the heating of swimming pools, which require lower water temperatures than a domestic supply.

One of the possible reasons for the Government's allocation of finance to develop solar systems is to improve the standards of construction and installation. The solar panel business is experiencing a boom similar to that of double glazing in the late 1960s. Unfortunately some enterprises have either misled the public with over-extravagant advertising claims or their standard of workmanship has left much to be desired; both issues have been the subject of many complaints to consumer associations, to the embarrassment of ISES.

In terms of cost-effectiveness, solar panels compare with double glazing, which has the poorest return on investment of all methods of thermal insulation.[16] The Department of Energy envisages that market penetration will be slow until insulation measures are carried out on a large scale. The Department argues that costs would have to be reduced by a factor of 3 or fuel costs would have to increase by an equivalent amount to justify the large-scale development of solar water heaters. ISES and the Select Committee, on the other hand, would prefer early investment. ISES feel that the Treasury's 10 per cent discount rate – the basis of all Government cost-effectiveness estimations – is false when the current rate of inflation and the nature of the competing fuels is considered, in particular the relatively short life of fossil fuels compared with the inexhaustible supplies of solar energy.[17] The Select Committee indirectly endorses this claim in its advocacy of a 50 per cent grant towards the installation and capital costs of a domestic water heating system up to a maximum of £400.[18] It is doubtful if the Government will accept this recommendation in the short term as it is presently financing schemes to reduce the cost of collectors by simplifying the design to facilitate mass production and installation.

The Government could encourage local authorities to incorporate solar panels into the design of new houses and in the rehabilitation of other property, thereby testing the performance of collectors. Schemes are in operation throughout the country, in some cases incorporating space and water heating (Lewisham) and in others only water heating systems (Edinburgh). The Edinburgh project is part of the Fountainbridge Housing Association's rehabilitation programme where a four-floor tenement block – with an advantageous roof position – is being converted to include a solar water heating system. The lack of response by most councils to solar heating experiments is related to the structure of council housing costs. The Housing Subsidies Act 1967 mainly considers capital costs of any scheme and extra capital costs, however meritorious, are difficult to justify regardless of the running costs involved.

Solar space heating systems

'The collector area required to meet mid-winter domestic heating demands using only the winter solar input would be many times the area of the roof.'[19] This observation by the Department of Energy highlights the major drawback to the development of solar heating systems; that of thermal storage. Solar radiation levels are greatest in the summer months when the heat is not required. As a result, this heat needs to be stored for later use and a house has to be designed to provide considerable heat storage capacity. The problem is not unique to the UK. Even in more climatically favoured countries such as the USA or Japan a heat store is necessary to maintain indoor temperatures at night or during overcast weather conditions. Water, rock and pebbles are the most common materials in use for storage systems. Water stores most heat per unit volume but costs are higher because the container must be watertight. The Wilsons in their three-bedroomed home at Long Sutton, Hampshire, have 6,500 gallons of water in plastic bags underneath the foundations.[20] Rocks and pebbles are also cheap materials and although their specific heat is lower than that of water, the storage area is smaller and does not have to be watertight. Considerable research is being undertaken into the use of chemicals for heat storage, which would further reduce the area occupied by storage in the house. The chemicals required would absorb heat at temperatures similar to those generated in a solar collector and would regenerate this heat as temperatures fell. The two chemicals which have been studied in experimental solar houses are sodium sulphate (Glauber's salts) and sodium phosphate dodecahydrate, and the Energy Research and Development Administration (ERDA) in the USA is studying the use of hydrogen in this reversible chemical reaction.

Until the problem of thermal storage is adequately resolved, solar systems will make little impact on the space heating market in the UK. The experimental solar houses depend on the use of the heat pump for efficient heat management,[21] but this and other modifications to conventional house designs – storage space, solar panels and extra insulation – make costs prohibitively high for the commercial market. It is not surprising, therefore, that many of the publicised solar homes are of the do-it-yourself variety.[22] However, Wates, the builders, are building solar homes in the London area. Their three-bedroomed prototype constructed at the National Centre of Alternative Technology demonstrates the benefits of the heat pump in association with a high standard of thermal insulation – 'double double glazing' and 18 inches of insulation in the roof, walls and under the ground floor. The house costs £22,000 to build including £3,000 for the low-energy refinements, although the commercial versions will be cheaper as a result of a reduction in insulation levels. Solar space heating systems could secure market penetration as fossil fuels increase in price and experience is gained in solar panel and heat pump technologies. However, the key to solar house development lies in a breakthrough in heat storage technology.

Solar electricity

The direct conversion of solar energy into electricity is an attractive proposition in that this power can either be fed into the National Grid or it can be used on a small-scale community basis. The use of solar cells for electricity generation provides the best possibility for development in the UK. The process has been known for over 40 years, from its original photographic applications to its development in space technology. In space programmes the cost of solar cells is not of great importance in relation to total costs, but for commercial electricity production, costs are too high by a factor of 100.[23] Basic material preparation accounts for 46 per cent of the cost of a silicon cell (Table 5.3). The refining of the silicon, especially the cutting of the crystal before polishing, is attracting most attention as a method of cost reduction. Prices have fallen by 15 times in 15 years, and comparisons have been made with price reductions achieved in the production of transistors. The PA Technology Centre in Royston claim to have made the manufacturing breakthrough.[24]

Despite the Department of Energy's lack of enthusiasm, solar cells could make an important contribution to UK energy requirements by the end of the century. They are pollution-free, contain no moving parts and require little maintenance. An additional bonus is their flexibility of use – the number of cells can vary according to the amount of power required.

Table 5.3 Cost breakdown of a silicon solar cell (%) (*Source:* Department of Energy, *Energy Paper No. 16*)

Purified silicon compound	11
Pure silicon	11
Single crystal growth	24
Cell manufacturing	34
Array assembly	24

Storing electricity, however, is as much of a problem as storing heat. The only practical method in use is conventional batteries, which are both bulky and expensive. At present the world's two largest oil companies are trying to produce a suitable fuel cell electrode at an acceptable price. Williams states that lead acid batteries weigh 50 lbs for every kilowatt/hour stored, whereas the methanol-air fuel battery, the possible battery of the future, uses three-quarters of a lb of fuel for every kilowatt/hour generated.[25] An alternative method is the electrolysis of water with surplus electricity to produce hydrogen, which is easier to store and transport. Indeed, the Japanese intend to move away from an oil-based economy to the large scale use of hydrogen in the future.[26]

Electricity can be generated from solar energy by other methods which are not applicable to the UK because of a lack of direct solar radiation. Solar power stations are being developed in the USA, France and Southern Italy, in the last as part of an EEC research project. These power stations operate on the 'power tower' principle, in which the sun's rays are focused on an array of mirrors. The reflected sunlight is concentrated upon a receiving tower which converts water into steam to generate electricity. It is possible that the UK could develop technologies which might be used in sunnier climates and ISES suggests that solar pumps and engines should be developed in Britain for export abroad. On a more grandiose scale the National Aeronautics and Space Administration (NASA) hopes to harness solar energy in an environment which receives uninterrupted sunshine every day of the year – outer space. With the assistance of the American space shuttle, NASA intend to assemble a solar collector at an altitude of 22,000 miles, where solar radiation would be converted to microwave energy before being beamed down to earth. Among the host of problems, apart from the cost, related to this project is the amount of energy required to initiate such a scheme.

Biomass

One aspect of solar energy which is often overlooked is the amount of energy stored in biomass – organic materials in the form of crops or organic wastes.

The poor efficiency of photosynthesis (0.16 per cent for all crops grown in the UK) would necessitate large areas of land to collect this energy. In Britain we produce sufficient biomass annually – the equivalent of 12 per cent of our primary energy demand – to make a considerable contribution to our energy needs.[27] Much of the food grown in the UK is consumed by animals, hence neither is our agricultural system geared to the production of energy, nor do we have a surplus of good land to devote to energy production. The Department of Energy recommends a development strategy based on the utilisation of agriculture and domestic organic wastes for the production of energy. The potential of such schemes will be discussed in Chapter 6, but clearly there is adequate scope to improve the energy efficiency of our existing agricultural system before any radical modifications are made to transform the farm economy with energy crops. Nevertheless high-yielding energy crops can be grown in marginal areas which are unsuitable for conventional crop production. The water hyacinth is an appropriate example; the plant has a dual function of producing methane at a rate of 6 cubic feet per pound, whilst also being used to purify water systems.[28]

The conversion of biomass into fuel is a well-established technology – methane production from anaerobic digestion is the most common method. During the Second World War a shortage of fuel led to the rapid development of methane plants in Britain and the Continent, but with the return of peacetime conditions interest waned in Europe, though the process was adopted by countries such as India, with more favourable climatic conditions and appropriate agricultural systems.

In northern climates such as that of the UK additional heat supplies are necessary to maintain the temperature of the digester to allow decomposition to occur for maximum methane production. The transport costs involved in the collection of the raw materials – sewage or agricultural wastes – and the further disposal of sludge fertiliser after the gas is produced, precludes the development at present of such schemes on economic grounds. Methane production from sewerage works for instance, necessitates the transportion of sewage from cities and, after production, the redistribution of the fertiliser to rural areas. Although successful methane production at sewerage works is practised in Leicester and Cambridge, high capital costs in addition to the transport costs will postpone the development of methane plants until fuel shortages promote a greater sense of urgency about the more efficient utilisation of our energy resources. Vale and Vale feel that mixed farms would benefit from a digester system because gas would be produced for cooking, and the treated sludge could be applied to the soil.[29]

Tidal power

The Seven Estuary has one of the highest tidal ranges in the world and the power potential from the tides could satisfy 12 per cent of the electricity requirement of England and Wales.[30] The idea of a barrage to harness this energy is not new; in 1933 the Severn Barrage Committee proposed a barrage costing £38.5 million, which would have produced 8 per cent of the country's power by 1941. The Germans had similar plans for the estuary if an invasion had been successful. Interest waned, however, until the publication in 1970 of the Underwood-Snow plan which envisaged a double-basin scheme, in which electricity could be generated continuously, with water passing from the sea to the high basin, from the high to the low basin and the low basin to the sea. The increase in the price of fuels in the 1970s created renewed interest in the energy potential of this site. The 1970s proposals were on a grander scale than those of 1933; they involved a barrage across the Severn between Cardiff and Weston-Super-Mare, instead of the initial scheme to construct a river barrage near the Wye confluence. Costs, and technological and environmental problems, are therefore also of greater magnitude.

Unfortunately the proposers of the feasibility studies have produced differing results. In 1974 Hughes and Glanville estimated that the capital cost involved (£2,000 million plus or minus £500 million) was competitive with that of conventional generating plant.[31] In 1977 the Netherlands Engineering Consultants Foundation (NEDECO) raised these figures to $3,000 million for a single-basin and £4,000 million for a bouble-basin scheme.[32] At the same time the report presented by the Hydraulics Research Station (HRS) drew a different conclusion from NEDECO about the impact of the barrage on the tidal range of the river, predicting an increase of 1.4 metres compared with NEDECO's estimate of a decrease of 1 metre in the range. These marked differences of opinion have prompted the Select Committee on Science and Technology to suggest that one independent body should be responsible for all future work on the feasibility of a barrage scheme.[33] The Select Committee strongly criticised the Department of Energy for its timid approach towards tidal power, claiming that the Government has not made serious efforts to evaluate its feasibility. The Department has been cautious in its approach, unconvinced that £4,000 million of Government money should be committed to a technically difficult scheme which could take 15 to 20 years to build, and which would produce only 10 million tons of coal equivalent per year.[34] The Government stated that this would be a 'one-off' project and the experience gained from its construction could not be utilised elsewhere. No small-scale prototype could be successfully built and therefore all technical and economic resources would be committed to one basic decision to construct. 'We are as yet nowhere near such a decision.'[35]

The Rance estuary scheme in France and a very small Russian project are the only operational tidal power plants in the world, but the Rance barrage is one-thirtieth of the scale of the proposed double-basin scheme. The conditions and method of construction would pose much greater problems in the Severn Estuary, to such an extent that the French experience cannot be related to the Severn project. On the other hand, the Select Committee questions the Government's open preference for wave power, which has yet to be proved on a significant scale and lacks the engineering techniques already established in barrage construction.

The problems of undertaking a venture of this magnitude cannot be resolved easily, if at all. Along the north-east Atlantic coast, the Canadians have been facing similar problems. Indeed, the history of the development of tidal power schemes has followed a similar course in the UK and Canada. In 1956 the US and Canadian Governments formed a joint commission to undertake a feasibility study of the tidal power potential of the Bay of Fundy (Passamaquoddy Project). In 1966 the Atlantic Tidal Power Programming Board was established with a similar brief in Canada.

Both reports, which appeared three years after the establishment of their respective organisations, rejected immediate development because of the high capital costs involved. Nevertheless, the changing fuel situation in the 1970s has created a revival of interest and in 1975 a further feasibility study was undertaken. Initial results have been promising enough to encourage the authorisation by the Nova Scotia Government of the construction of a small 10–20 MW test generating unit on the River Annapolis, whilst parallel studies are conducted into the possible socio-economic and environmental effects of barrage construction.

On both sides of the Atlantic, the harnessing of tidal energy would create ecological imbalance, although the extent of the environmental effects would vary according to the local conditions experienced in the Severn Estuary and the Bay of Fundy. The construction of a power plant would alter current patterns, reduce tidal velocity and exchange (flushing), lessen salinity in upper layers and lead to higher water temperatures in summer, and possible winter icing. The main overall physical effect would be the alteration of sedimentation and erosion patterns. In the Bay of Fundy and the Severn Estuary, large amounts of sediment, mobile under the influence of strong tidal currents, are deposited on the seaward side of the estuaries, and a barrage would result in silt accumulation upstream. In the Severn Estuary the heavy glacial erosion of the Welsh valleys has resulted in sediment accumulation which is removed by tidal scouring and therefore a more substantial problem is posed than at the Canadian site. The disruption of fish migration and breeding patterns would be a serious problem in the Bay of Fundy, but Wilson states that in the Severn Estuary the barrage would have little effect on the fishery, basing his judgement on the breeding habits of the salmon in Scotland, where fish have to ascend and descend faster turbines than those envisaged in the Severn project.[36]

The main environmental problem associated with the Severn project is the danger of aggravating the high pollution levels which presently exist in the estuary. The Passamaquoddy scheme anticipated no such problems because of the small amount of pollution entering the project area;[37] the 1969 Fundy report notes that inadequate tidal flushing in parts of Prince Edward Island had led to seasonal stagnation and pollution.[38] The report adds that this situation would be worsened if the new power source attracted population and industry to the region. In the Severn Estuary the problem is much more acute. The estuary suffers from heavy metal pollution from industry at Avonmouth and has the undistinguished record of receiving the greatest volume of crude sewage – 47 million gallons per day – of any estuary in England and Wales.[39] Clearly, the flushing action of the tides helps to disperse, dilute and remove this pollution, but on impoundment slower tidal velocities and reduced dissolved oxygen in lower layers would intensify pollution, eventually culminating in the extermination of all marine life. If the scheme is approved, pollution control measures to purify the estuary would be essential.

Fortunately, the construction of a barrage also has advantages. In all the schemes under review, the tourism and recreation potential, the improved communications network and the availability of power could give a spur to regional development. The Rance scheme utilises the top of the dam as a highway across the estuary and has also become a tourist attraction. Severnside has experienced growth in a period of national recession; the improvement of communications from the South-East has attracted industrial development to Avonmouth, and Bristol is currently one of the fastest growing provincial office centres in Britain. The barrage could enhance this growth potential and have wide regional implications for the future industrial and urban development of South Wales. Although barrages could provide navigational problems, Brown anticipates that a barrage across the Severn could provide *improved* port facilities with the prospect of 250,000-ton vessels gaining access to Bristol docks given the provision of suitable locking facilities.[40] An additional recreational advantage would be the possibility of using the calm waters behind the barrage for sailing; at present the high tidal range makes this impossible.

In spite of the advantages of a barrage scheme, a decision on its construction is several years away. HRS and NEDECO cannot agree on the effects the barrage will have on the tidal range and therefore it is clear that much more research is necessary to achieve a complete understanding of *all* the possible effects of barrage construction. Until some of the ecological uncertainties are removed, it would be unwise to build a barrage across the Severn Estuary. P. H. Waller, an engineer, writing in 1970, summarises the situation which is still applicable to the Severn project at the time of writing.

Uncertainty about ecological effects, and the haunting knowledge of recent examples of unforeseen and unwanted spin-offs from technological changes, argue for careful consideration of such effects in the case of large scale, capital intensive, and essentially irreversible schemes such as tidal power developments. In the absence of certain knowledge of effects, the course of action that preserves the option to reverse a decision in the light of new knowledge has an inherent advantage over a course of action that is irreversible. Making of an irrevocable decision carries with it a heavy responsibility to assure that every effort has been made to anticipate all effects of that action, including effects that may be subtle or remote.[41]

Geothermal power

Geothermal power is not a form of energy usually associated with the United Kingdom. Situated at a distance from active tectonic plate boundaries, the long-term stability of the British land mass offers little potential for capturing 'high enthalpy' power from hot acquifers as has been done in Italy, New Zealand, Iceland and the USA. Steam in reservoirs of this type reaches temperatures between 200° and 350°C and the heat recovered or electricity generated plays an important role in the energy supply of these areas.

More relevant to the British situation is the concept of 'hot rock' or 'low enthalpy' power, because these geothermal conditions are more widely represented in Britain than 'high enthalpy' conditions.[42] The warm springs of the Bath-Bristol region have been used for therapeutic purposes since Roman times. The discharge water temperature – 48 °C – is too low for power generation purposes; however, the reservoir is still unexplored and the Department of Energy estimates that the temperature could exceed 100 °C at the source. Buxton, a Pennine spa town, possesses hot therapeutic waters which percolate through the carboniferous limestone. Much of the power potential of the Pennine region is limited because of low water temperatures – a maximum of 60 °C. The other geological areas where rainwater percolates to sufficient depth for its temperature to reach 100 °C are sedimentary basins, for example, the Cheshire and Hampshire basins.

The 'hot rock' concept depends on rocks that have a higher than normal temperature because of local heat sources but lack permeability to allow natural water circulation. To release its energy the rock is drilled and then cracked at the base of the hole to allow an adequate heat transfer area. Water is then pumped through the system and recovered through a second hole. The areas of greatest potential for 'hot rock' development are two granite baccoliths, one of 1500 sq. km at Durham, the other, four times the size, in Cornwall.

The Government initially backed the Energy Technology Support Unit (ETSU) with £250,000 to investigate the possibilities of tapping the earth's natural heat. The ensuing report the following year outlined the Department of Energy's case for research into this energy source and the Government allocated a further £840,000 to ETSU to continue the investigation of geothermal potential in the UK.[43] By early 1978 some interesting discoveries had been recorded. Scientists working for the Department of Energy have made two new hot water 'strikes' which show commercial possibilities. The water from a strike 10,000 feet below the coastal strip near Hull could be piped to some of the city's 6,500 council flats, or to industrial processing plants, whilst the water from a discovery in the Scottish central belt could find a market in either the Clydeside or Forth urban areas.[44]

The extent to which potential geothermal sites will be exploited depends on their competitiveness with fossil fuels. High capital costs – for example, an estimated £500,000 for one borehole in North Humberside[45] – necessitate the minimisation of distribution costs. It is advantageous that discoveries be located in areas which require low grade heat in order to minimise both heat loss and additional costs which would be incurred through piping the water over a long distance.

Geothermal heat seems unlikely to make an important contribution to the country's energy budget by the turn of the century. Nevertheless, the national demand for low-grade heat is not insignificant, and an impact could be made in regional 'pockets'. The North Sea has some of the highest temperature gradients recorded in the UK with deep sections of sedimentary rocks recording temperatures in the range of 150–200 °C. When oil and gas reserves become depleted, drilling technology could be utilised conveniently to exploit the country's area of highest geothermal potential. By 2020 the redundant infrastructure of the oil and gas industry in the North Sea may be producing geothermal power for use in eastern Britain.

Wave power

The Department of Energy has shown more enthusiasm for wave power than for any other form of renewable energy resource, and the Government has backed this interest with financial support for the five most important wave power projects. Interest in this technology has a long history; between 1856 and 1973 more than 340 wave energy systems were patented in Britain.[46]

The inherent attraction of wave energy is that it would be available throughout the year and peak output would coincide with the pattern of peak electricity demand (unlike the case of solar energy). An open sea-front with a long uninterrupted fetch and prevailing onshore winds provide optimal conditions for wave power development. North Atlantic waves cover a long fetch of over 500 kilometres in consistent, regular wind conditions which provide excellent potential for the development of the north and west coasts of Scotland.[47]

Although Salter's 'ducks' have received most publicity, other wave power devices are being considered by the Department of Energy. Sir Christopher Cockerell has devised a scheme of connected rafts which lie at right angles to the waves. As the rafts move and tilt, piston-operated pumps at each joint convert the mechanical energy created into power. This scheme is at the small-scale model stage and the full-size plant would have rafts 10 metres long and 20 to 40 metres wide.

The National Engineering Laboratory is developing a wave-power device based on a principle devised by the Japanese engineer Masuda from his research on breakwaters. By dividing a buoy into a large number of chambers, energy can be produced through a low-pressure air turbine when air is displaced by the rhythmic action of the waves. Like most wave-power devices the full-scale apparatus will be spectacularly large – the buoy may be up to 300 metres in diameter.

The other two devices receiving Government support are the Russell rectifier and French's 'inflatable sausages'. The team working under Professor French at Lancaster University believe that their system is cheaper to build than the other devices. It consists of a long inflated 'sausage', divided into compartments with a series of high-pressure mains suspended beneath the apparatus. Waves crossing the 'sausage' squash the inflated tube, pushing out air into a pressure main which drives a turbine connected to a generator. When the wave moves on, the pressure in the sausages falls and air flows back into it from another pressure main. The rectifier was named after the Director of the Hydraulics Research Station, Robert Russell. The rectifier is split into two compartments – an upper and lower reservoir. Water flows into the upper level as waves flow over the device, creating a head of water to drive a turbine as it empties into the lower basin.[48]

All of these projects are still at the prototype stage. The most promising and most publicised wave energy project is Stephen Salter's 'nodding duck' generators. Development has moved on from the 'bath tub' stage to the testing of the performance of the ducks by Lanchester Polytechnic engineers at Draycote reservoir near Rugby. Further experimentation is being carried out at Loch Ness on a one-tenth scale model, before a large-scale prototype is constructed offshore near Dounreay. Salter has shown that the original research idea required very little financial support, and much of the basic design testing was achieved on a shoe-string budget. With an original budget of a few hundred pounds, Salter modified the shape of the generator from a vertical vane pivoted round a horizontal axis, achieving 40 per cent efficiency, to the duck shape, which displaces less water astern, and increases efficiency to over 80 per cent.[49] On scaling up the project to £10 million wave-power plant units, the ducks will be between 10 and 20 metres in diameter, strung together over a distance of 500 metres. Because of the length of the backbone, an excessively large wave – 'the 50-year wave' – could result in a dangerous bending movement in the centre. This is the most important problem to be resolved, because the amount of steel required to strengthen this section could increase capital costs sufficiently to make this wave-power project uncompetitive with alternative energy generation methods. Salter believes that design modifications would overcome this problem, and if sea trials prove successful, wave-power plants could be producing commercial electricity by the 1990s.

Where would these plants be located? It was mentioned earlier that the north and west coasts of Scotland are potential locations because of their exposure to prevailing onshore winds and waves which cover a long, uninterrupted fetch. Additional locational requirements are deep-water sites (because of the size of the ducks) close to the shore to minimise transmission costs.

All of the best sites in Britain are distant from the main areas of demand. Nevertheless, Salter estimates that the power transferred from sea to duck, the conversion to electricity and transmission to the south-east of England would have an overall efficiency of 60 per cent.[50] At the time of writing the most optimistic forecasts for wave-power costs approximate to the highest estimates of fossil fuel generation costs, but its cost competitiveness should improve in the near future. If wave power becomes competitive in the next decade, other advantages could accrue from the establishment of wave generators in offshore Scotland. Britain's ailing shipbuilding industry might receive a much needed boost; the construction of 1000 MW of generating capacity would be the equivalent to building 60 500,000-ton supertankers.[51] Perhaps the nearby availability of power would revitalise some of those areas which have suffered from rural depopulation in the past. However, the development of wave-power plants would encounter difficulties. Running costs would be higher than for other renewable energy sources. Hydraulic parts would need replacement bearings and seals approximately every six years, and the main structure would require antifouling treatment at regular intervals. The unmanned generators could be a danger to navigation although the stationary installations would have marked chart positions in such a way that if an accident occurred and one unit broke free, it would sink rather than be a potential danger to shipping.

Two environmental effects of wave-power generation would be, firstly, the necessity for transmission pylons across areas of unspoiled countryside and, secondly, the ecological imbalance which would be created by removal of energy from the waves before they reach the shore. Lower wave velocities would cause changing erosion and sedimentation patterns, the precise effects of which are difficult to ascertain without a detailed environment impact study. Moreover Salter concedes that mistakes in siting could be made, though the power plant could be closed down or re-located, which is the greatest advantage of wave power over tidal power. Once the decision is taken to develop wave-power technology, the nature of the plant gives a degree of flexibility which could never be offered by a tidal power scheme.

Wind power

The wind has been a useful source of power in the past – 10,000 windmills are estimated to have been in use throughout Britain in the nineteenth century for a variety of purposes from grinding corn to pumping water. If wind power is to make a significant contribution to future UK electricity supplies, however, traditional technology must be modified, especially in terms of scale, with the picturesque, 'dutch' windmills being superseded by aerogenerators of 200 feet (70 metres) in diameter. The Department of Energy has shown little enthusiasm for wind power development, and in 1977 it was felt that wind power had similar potential to tidal energy, but less potential than wave power.[52] The announcement by the Government in January 1978 that a giant demonstration windmill would be constructed indicates a change in the official attitude towards wind power. The exhortations of experts such as Musgrove and Ryle may have influenced the Government to commit financial support to the development of wind energy systems. Ryle, a Nobel Prize winner, argues convincingly that wind energy is the best method of making a significant contribution to reducing the energy gap by the early 1990s, thus allowing the preservation of some of the North Sea oil for the future.[53] If electricity is to meet the space heating loads previously filled by oil and gas, the peaks in energy demand can be met more satisfactorily by wind than by nuclear generation. Ryle has demonstrated that wind-generated electricity can be stored more cheaply to meet peak demand than nuclear electricity. An additional advantage of wind systems over energy competitors is the short time required for a build-up to full-scale production.[54] If the Government were to encourage windmill construction Ryle contends that 2,000 units per annum could be produced within five years.

Which areas of the UK offer the most potential for wind development? In terms of mean annual wind speeds and percentage of annual calms, the best sites coincide with potential wave-power areas; for example, exposed sites in the Hebrides register sufficiently high wind speeds for aerogenerators to be competitive with conventional power stations (Fig. 5.1).[55] Nevertheless, aerogenerators do have some drawbacks. Wind-power systems sited in remote areas of north and west Scotland would experience high transmission losses, similar to those encountered in wave-power development. Moreover, the poor overall efficiency of aerogenerators in comparison with wave-power plant could eventually favour wave rather than wind power development. Environment considerations also might delay the construction of windmills in areas which are as beautiful as they are remote. The noise and visual intrusion of windmills, which would be bigger than pylons, would doubtless lead to much opposition (Fig. 5.2). Professor Lewis of Newcastle University argues that 60,000 electricity pylons have been visually accepted by the British public and that 6,000 windmills of a more pleasing design should present fewer aesthetic problems than the further construction of conventional power plants.[56] Some of the environmental objections could be overcome if aerogenerators were positioned offshore. Musgrove advocates this approach and both the Department of Energy and the Select Committee do not dismiss the possibility for the future.

Musgrove believes that the UK should follow the example of the Dutch, who have greater pressures on land use than in Britain and therefore intend to build windmills in shallow offshore waters.[57] Although wind speeds are greater offshore, the costs of offshore installations and transmission to land could outweigh this advantage. Nevertheless Musgrove maintained that improvements in windmill design and increases in fossil fuel costs had made wind power systems competitive with conventional power systems by 1977.[58]

At Reading University Musgrove's team are developing a vertical-axis windmill with an efficiency which compares with conventional helicopter-blade designs, but which is cheaper and less obtrusive. The original vertical-axis design was patented, but never used, by Georges Darrieus in the 1920s, and Musgrove's simplified H-shaped design (Fig. 5.2) has been patented by the National Research Development Corporation, which hopes to introduce the windmills on the commercial market. The three-metre prototype is intended to be scaled up to 70 metres in use and Musgrove believes that the best prospective site for the windmills is a shallow, windy area in the Wash (Fig. 5.3). The annual average wind speed in this region at a height of 50 metres above sea level – the centre of the windmill – is 9 metres per second and therefore a cluster of 400 windmills 0.5 kilometres apart would provide a power output of 1000 MW. Each cluster would occupy an area of 100 square kilometres, represented by the insert of Fig. 5.3. Within the Wash the best location for the deployment of windmills is more than 10 kilometres from the coast, to avoid calm zones, but no further than 50 kilometres offshore, because of high tranmission costs. Musgrove calculates that the area between the 10 to 50 kilometre zones which is in shallow water, less than 20 metres in depth, exceeds 4000 square kilometres. If only one-half of this area was zoned for wind-power generation, it could supply one-third of our present electricity requirements.

Musgrove argues on similar lines to Ryle for the development of wind power systems. The availability of wind energy coincides with peak demand when power stations low in the merit order come into operation; hence, he claims that windmill systems could be used as fuel savers. An additional advantage of

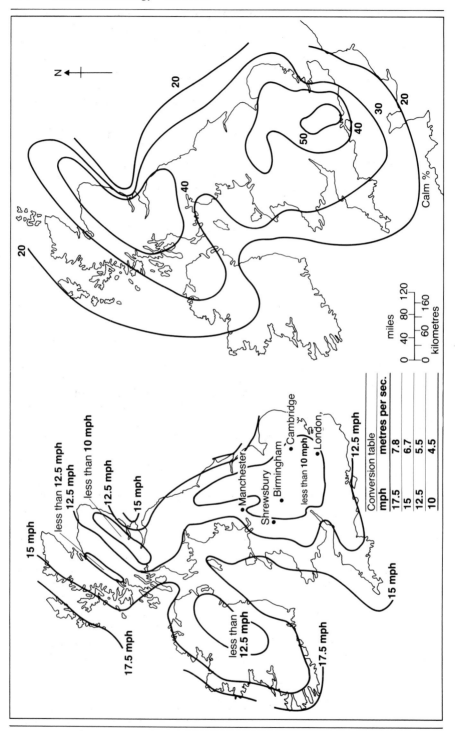

the southern North Sea as a prospective wind power area is that nearby depleted gas reservoirs could be used for energy storage. In 1977 Musgrove estimated that an offshore windmill system cost £580 per kilowatt – a figure which compares favourably with nuclear plant costs (Table 5.4). The practical implementation of Musgrove's ideas will produce problems, although not insurmountable problems. He suggests that the area earmarked for development could be zoned into sea use areas to avoid conflict between competing users.

The environmental effects of large-scale offshore windmill developments have not been studied in detail, but Musgrove doubts if there would be any noticeable effect on coastal ecology because of the distance the windmills would be deployed offshore.[59] Indeed, he feels that windmill clusters would provide a safe breeding area for fish, as large fishing vessels would be prevented from entering this zone.

Table 5.4 Costs of an offshore wind energy system (£/kW) (*Source:* P. Musgrove, *North Sea Wind Energy,* 1977)

	£/kW
Conventional windmill	280
Offshore siting	100
Submarine cables (max. of 50 km)	50
Compressed-air storage plant	150

The advantage of wind power over some of its renewable energy competitors is its flexibility of scale. Musgrove has stressed the importance of his small-scale windmill for use on farms and in remote areas. The Department of Energy reiterates this point, adding that further research into the development of small aerogenerators (under 100 kW) could reveal their export potential.[60]

Some experimental wind-power sites have already been developed; for example, at Peacehaven in Sussex the first council houses in Britain to be supplied by wind power are being constructed on a cliff-top site.[61] Britain is too urbanised to make 'a windmill on every roof' a practical possibility. Consider the legal and planning ramifications when house-holders began to claim that their wind was being stolen by another aerogenerator constructed on a nearby roof! These problems would not occur in rural areas, and these therefore offer the best possibilities for small-scale wind power developments. The increase in price of fossil fuels has stimulated interest in the agricultural sector, where energy accounts for a large proportion of total costs. Growers estimate that the cost of heating an acre of tomatoes for one season totals £11,500, whilst the pig production industry has incurred increased costs in maintaining temperatures of over 20 °C in weaner buildings.[62] Wind power is an attractive proposition to farmers who are presently dependent on large inputs of oil or gas, and the performance of a trial windmill and glasshouse heating system in Hampshire is of particular interest to high energy users. At the experimental horticultural

Fig. 5.1 Isovents showing mean annual wind speed in the British Isles and per cent annual calm (including frequencies of winds up to 6 knots) (*Source:* B. and R. Vale: *Autonomous House,* Thames and Hudson 1975; and Department of Energy, *Energy Paper No. 21,* HMSO 1977)

Sizes of aerogenerators compared with grid pylons
(A) Traditional marsh drainage windmill, Soham Mere, Cambridgeshire
(B) CEGB Suspension tower – CEB L 132 (1940) 132 kV double-circuit
(C) CEGB Suspension tower – BES L 3 (1956) 275 kV double-circuit
(D) CEGB Suspension tower – BEBS L 6 (1966) 400 kV double-circuit
(E) 1 MW rated (28 knots) ETSU/Servotec aerogenerator

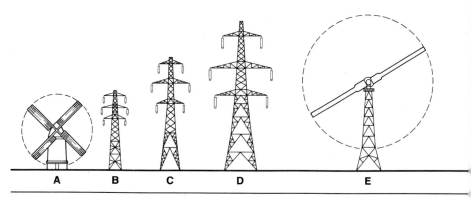

Fig. 5.2 Scale of aerogenerators (*Source:* Department of Energy, *Energy Paper No. 21,* HMSO 1977)

station at Efford it is hoped that considerable economy will be achieved with the introduction of wind energy to supplement the existing oil heating system in two quarter-acre greenhouses.[63]

Ocean Thermal Energy Conversion (OTEC)

The oceans store and collect vast quantities of low-grade heat energy which can be converted into electricity. The process was invented by d'Arsonval as early as 1881, and in 1930 another Frenchman, Claude, showed experimentally that heat from surface water could convert either sea water or a working fluid such as ammonia into a gas to generate electricity; the gas in turn is condensed by pumping cold water from depth.[64] This refrigeration cycle operated at sea is being explored in the United States by the Energy Research and Development Administration (ERDA), which hopes to test a prototype plant by the early 1980s.[65] The best sites for these plants are in zones where the surface water is consistently warm with a steep offshore gradient to enable cold water to be pumped to the plant. Proximity to the shore and market would also minimise transmission costs. American interest lies in the possibility of using the Florida current to provide power for the eastern seaboard. British interest lies not in the availability of power near her shores but in the provision of OTEC plants to export markets. The semi-submersible plants will be on a similar scale to the large oil and gas production platforms already in operation in the North Sea. The experience of designing and constructing plant of this scale for an offshore market could resuscitate British platform yards, which suffered from fluctuating demand in the mid-1970s.

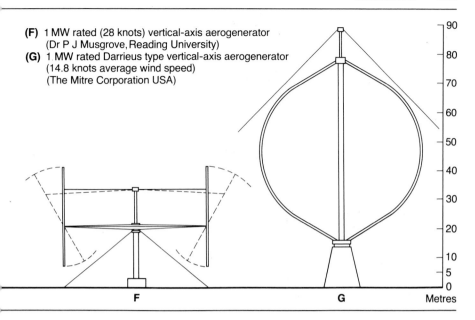

(F) 1 MW rated (28 knots) vertical-axis aerogenerator
(Dr P J Musgrove, Reading University)
(G) 1 MW rated Darrieus type vertical-axis aerogenerator
(14.8 knots average wind speed)
(The Mitre Corporation USA)

Conclusion

The 'energy crisis' of 1973/74 did stimulate a renewed interest in established renewable energy technologies which could have commercial possibilities in the near future. With North Sea oil and gas production well under way in the late 1970s, renewable energy resources need to be developed for the future. The Government announced increased support for developments in this area in 1978. "If all the projections of energy supplied from schemes outlined in this chapter were realised, renewable energy resources could supply all of Britain's electrical energy by the first quarter of the next century. Even if the unlikely assumption is made that Britain will adopt a low-growth, non-nuclear energy strategy for the future, the Government would only encourage a transitional move to renewable resource development whilst existing nuclear plants repaid their construction costs. As a result, renewable energy resources will be developed according to their economic and technical potential over a period of time. Wind and wave power show the best possibilities. Wind-power technology is at an advanced stage and production could be built up quickly; wave power, on the other hand, offers the prospect of greater quantities of power at higher efficiency and employment opportunities in existing shipyards, but the scaling up of prototypes will take much longer than for wind power. Geothermal energy is more of a long-term prospect although geographically favoured areas could utilise the earth's heat for district heating purposes by the 1990s. A question mark hangs over the viability of constructing a barrage across the Severn Estuary to harness the considerable energy from the high tidal range. The main problems here are not only economic and technological, but also environmental as the already high levels of pollution and sedimentation will be aggravated by barrage construction.

More promising developments should occur in the field of solar energy as a

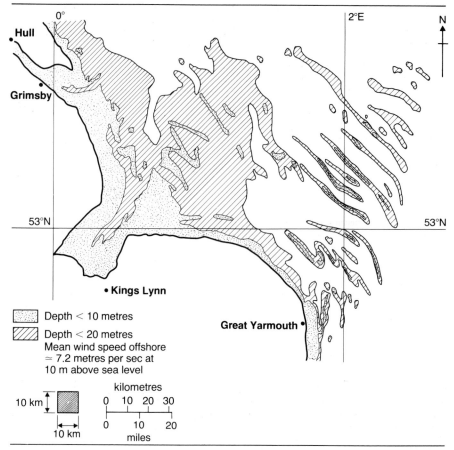

Fig.   **5.3** Shallow waters of the Wash (*Source:* After P. Musgrove , *North Sea Wind Energy,* 1977)

result of design improvements of solar water-heating systems whilst technological breakthroughs in solar cell production could lead to direct electricity generation in large or small units. This flexibility of scale in both solar and wind energy systems will ensure their continued development in the future. The systems complement each other, in that maximum energy production occurs at different times throughout the year and small-scale systems will increasingly penetrate agricultural markets where energy accounts for a high proportion of total costs.

Notes and references

1. Department of Energy (1977a) *Tidal Power Barrages in the Severn Estuary, Recent Evidence on their Feasibility, Energy Paper No. 23,* HMSO, p. 1.
2. The official bodies refer to 'solar energy' as the utilisation of solar radiation for heat or electricity generation and omit other renewable resources which are a part of solar energy systems. In the text 'solar energy' will be used in its 'official' context.

3. D. Dickson (1974) *Alternative Technology and the Politics of Technical Change*, Fontana.

4. G. Boyle and P. Harper (1976) *Radical Technology*, Wildwood House.

5. Select Committee on Science and Technology, Third Report (1977a) *The Development of Alternative Sources of Energy for the United Kingdom*, vol. 1, HMSO, para. 128.

6. S. Salter feels that the level of Government spending is about right, (personal communication, February 1978); see also D. Fishlock, 'Hidden costs of sun power', *Financial Times*, 14 May 1976.

7. Centre for Alternative Technology. *Visitors Guide*, p. 3. See D. McGuigan (1978) *Small-scale Water Power*, Prism Press. An example of small-scale water power in operation at Launceston, Cornwall, is illustrated in P. Dodd and G. Harvey (1977) 'Water, sun and wind, nature's power stores', *Farmer's Weekly*, 11 March, pp. xxi–xxiv.

8. Select Committee on Science and Technology (1977) op. cit., para. 129.

9. Compare Department of Energy (1976a) *Energy Research and Development in the United Kingdom*, Energy Paper No. 11, HMSO, p. 59 with *Solar Energy: its potential contribution within the United Kingdom*, Energy Paper No. 16, HMSO (1976b) p. 1.

10. International Solar Energy Society: UK section (1976) *Solar Energy: a UK Assessment*, ISES, p. 335.

11. Ibid, p. 12.

12. Ibid, p. 6.

13. For a general discussion see Alice Coleman (1977) 'Land use planning, success or failure? *Architects' Journal*, 19 January 1977, or a detailed case study, P. J. Smith (ed.) (1975), *The Politics of Physical Resources*, Ch. 5, Open University Press.

14. B. Vale and R. Vale (1975) *Autonomous House, Planning for Self Sufficiency*, Thames and Hudson, p. 19.

15. Ibid, pp. 60, 61.

16. Department of Energy, (1976b) *Solar Energy: its potential contribution within the United Kingdom*, Energy Paper No. 16, pp. 30–33.

17. ISES (1976) op. cit., pp. 28–33.

18. Select Committee on Science and Technology (1977a) op. cit., para. 144.

19. Department of Energy (1976b) op. cit., p. 33.

20. *Sunday Times*, 6 February 1977.

21. The heat pump is discussed in greater detail in Ch. 6 as a means of saving energy.

22. In the case of Granada TV's 'house of the future' at Macclesfield the high capital costs for the hardware were absorbed by the manufacturers for free advertising. See J. Willoughby, 'Good day Sunshine', *Undercurrents*, No. 23, September 1977.

23. There is some difference of opinion on solar cell costs; the Department of Energy claim a reduction in costs by 1,000 is necessary whilst the Select Committee and ISES quote a factor of 100.

24. Shown on BBC Horizon programme, *Dawn of the Solar Age*, 18 March 1977.

25. K. Williams (1977) 'The Ultimate auxiliary', *Yachting Monthly*, December 1977, p. 1886.

26. BBC, op. cit.

27. Department of Energy (1976b) op. cit., p. 5.

28. BBC, op. cit.

29. Vale and Vale (1975) op. cit., pp. 110 and 111.

30. B. J. H. Brown (1976) 'Tidal power from the Severn Estuary?', *Area*, **8**. p. 114.

31. Ibid.

32. Department of Energy (1977) op. cit., p. 11.

33. Select Committee on Science and Technology, Fourth Report (1977b) *The Exploitation of Tidal Power in the Severn Estuary*, HMSO, para. 59.

34. Department of Energy (1976a) op. cit., p. 62.

35. Department of Energy (1977) op. cit., p. 2.

36. E. M. Wilson, et al *The Bristol Channel Barrage Project*, Paper 103, 11th Conference on Coastal Engineering, London, 1968, quoted in D. H. Waller (1970) 'Environmental effects of Tidal Power Development', paper presented to the International Conference on the Utilisation of Tidal Power, Halifax, Nova Scotia, 1970, p. 12.

37. Ibid; p. 14.

38. Ibid; pp. 14, 15.

39. Department of the Environment (1974) *Report of a River Pollution Survey of England and Wales, 1973, vol. 3*, HMSO, p. 2.

40. B. J. H. Brown (1977) 'The planning challenge of tidal power', *Town and Country Planning*, March 1977, p. 162.

41. D. H. Waller (1970) op. cit., p. 20.

42. Department of Energy (1976a) op. cit., p. 58.

43. Department of Energy (1976c) *Geothermal Energy: The Case for Research in the United Kingdom*, *Energy Paper No 9*, HMSO. The Government has sought partial funding of the project from the European Commission.
44. *Yorkshire Post*, 15 November 1977.
45. Ibid.
46. M. Kenward 'Putting wild waves to work', *Sunday Telegraph Magazine*, 6 March 1977, p. 32.
47. S. M. Salter, D. C. Jeffrey and J. R. M. Taylor (1976) *Wave power – nodding duck wave energy extractors*, Paper presented at Energy from the Oceans Conference Raleigh, North Carolina, 27 and 28 January 1976.
48. Norwegian engineers, who have probably the most experience in marine engineering may have reached the most economical solution. Artificially creating an underwater earthquake produces rapid wave growth, giving high-speed breakers on reaching the shore. The capturing of this energy would be based on the 'rectifier' principle channelling water into a reservoir 100 metres higher than sea level.
49. S. H. Salter (1974) 'Wave power', *Nature* **249**, No. 5459, 21 June 1974, pp. 721, 722.
50. D. Mollison, O. P. Buneman and S. H. Salter (1976) 'Wave power availability in the NE Atlantic', *Nature* **263**, No. 5574, 16 September 1976, p. 224.
51. *Financial Times*, 17 June 1976.
52. Department of Energy (1977b) *The Prospects for the Generation of Electricity from Wind Energy in the United Kingdom*, *Energy Paper No. 21*, HMSO, p. 50.
53. M. Ryle (1977) 'Economics of alternative energy sources', *Nature* **267**, No. 5606, 12 May 1977, p. 117.
54. Ryle makes an analogy with the development of the hovercraft which flew eight months after the design work was started.
55. Department of Energy (1977b) op. cit., p. 49.
56. Personal communication, March 1978.
57. P. J. Musgrove (1976) 'Windmills change direction', *New Scientist*, 9 December 1976, p. 596.
58. P. J. Musgrove (1977) *North Sea Wind Energy*, a paper presented at a one-day conference, 'Do We Need More Nuclear Power Stations?', Huddersfield, 7 January 1977.
59. Personal communication, April 1977.
60. Department of Energy (1977b) op. cit., p. 50.
61. *Building Design*, 4 November 1977.
62. Dodd and Harvey (1977) op. cit., pp. xiii, xxxiii.
63. Ibid., p. xxxiii.
64. Robert Cohen (1975) 'Ocean Thermal Energy Conversion' in M. Granger Morgan (ed.) *Energy and Man*, IEEE Press.
65. Energy Research and Development Administration (1976) *Ocean Thermal Energy Conversion, Program Summary, October 1976*, ERDA.
66. On 6 June 1978, the Government announced support of £6 million for further research and development into renewable energy resources; the major share of the budget, £2.9 million, will be allocated to wave-power projects.

Chapter 6

Energy conservation

In the post-energy crisis era, governments initially enforced conservation measures upon their citizens because of a shortage of oil supplies. In Britain OPEC action was quickly followed by the miners' strike of February 1974, and in a matter of months the British public began to appreciate the need for energy conservation. A three-day working week, the curtailment of TV viewing beyond 10.30 p.m. and the prospect of petrol rationing highlighted our dependence on energy. Nevertheless, five years later, with North Sea oil coming ashore in greater quantities each month and after a series of mild winters and hot summers, the sense of urgency has perhaps begun to disappear. At the moment Britain is perhaps too well endowed with energy resources to appreciate that a barrel saved is better than a barrel won. Indeed, in the case of North Sea oil development, costs (including energy costs) in the manufacturing of steel and associated components to build rigs, platforms and pipelines undermine the energy value of the fuel produced.

Although energy conservation is envisaged as the cornerstone of UK energy policy in the future,[1] the radical transformation from a society geared to cheap energy to one based on energy thrift can only be achieved over a long period. Energy efficiency has to compete with cost efficiency and the Government and consumer alike remain complacent about energy waste and seem more concerned with the more immediate consideration of adequate returns on investment. Porteous cites the case of energy-prodigal office buildings in London where energy costs are such a small fraction of total costs that there is little financial incentive to introduce conservation measures.[2] The Department of Energy's discussion document on district heating combined with electricity generation states that the coal equivalent of 20 million tons a year could be saved if one quarter of UK housing was converted to district heating; however, the scheme was considered to be uneconomic given the current availability of competing fuels.[3] Chapman typified the attitude of the individual consumer when he made his decision to choose electric storage heaters in 1969 in preference to the more energy efficient gas or oil-fired systems.[4] His reasons were essentially economic – he could not afford the capital outlay for the alternative systems. All decision-makers – the individual, the local authority and the Government – have to analyse the potential returns on their investments. Home owners therefore plan for the most cost-effective energy systems related to the length of time they expect to remain in the same house rather than the best overall energy effective systems.

The energy users

The energy flow diagram (Fig. 6.1) illustrates the amount of energy wasted from the primary fuel source to its end use, and Fig. 6.2 shows the main energy consumers by sector. The elimination of wasteful practices and improvements in efficiency are generally cited as methods of saving energy. Unfortunately, the root cause of many of the UK's energy problems lies in the energy intensiveness of our modern economy and, in particular, in the move to automation and mechanisation which has led to the replacement of human energy by fossil fuel energy. This policy makes sense in terms of cost efficiency in a period of cheap energy supplies. Energy is however no longer cheap and its scarcity in the future will result in the energy input becoming a progressively more expensive item in industrial production costs. Undoubtedly society will adjust in the long term to less energy-intensive methods but in the immediate post-energy crisis period, no radical transformation has occurred in energy use.

Agriculture

The agriculture sector illustrates the cost effectiveness of substituting energy inputs for labour inputs to produce increasing yields from smaller acreages.[5] Although agriculture only consumes 1 per cent of total energy (Fig. 6.2), energy input per man is comparable with that of heavy industries. During the era of cheap energy supplies, 1952–72, agricultural energy consumption rose by 70 per cent, nearly twice as much as the increase in national fuel consumption during the same period.[6] Slesser claims that the UK population 'eats' 30 million tons of oil per year, although about half of this energy is utilised in the food processing industries which cater for the convenience shopper who forgoes quality for an expensive high energy-input product.[7] The food production sector absorbs five times as much energy as it yields in edible calorific value.[8]

Inevitably, the escalation in oil prices has led to improvements in agricultural methods to reduce energy costs. The horticulture sector in particular, is very dependent on the use of petroleum. Current research is devoted to the improvement of insulation in greenhouses through double glazing and the use of night blinds to reduce heat loss. The siting of greenhouses near energy-intensive industries or power stations, such as the experimental scheme at Eggborough in Yorkshire, opens up further possibilities in waste heat recovery. Savings could be made by agricultural fuel conversions. Straw could be processed as liquid or solid fuel instead of being burnt in the fields, or the heat derived from burning could be utilised to dry crops in place of the increasingly expensive electric methods currently used. Ecologists are keen to point to the savings that can be derived – in energy and environmental terms – from the substitution of organic for inorganic fertilisers. As only one half of the nitrogen fertilisers applied to the soil is consumed by crops, greater care should be taken in the timing of fertiliser application to minimise both the unproductive waste of energy and the environmental effects of polluting run-off.

In the long term, however, the production profile of British farming will inevitably change. Throughout the post-war period Government support policies, allied to an increase in demand from a more affluent society, increased food production from a self-sufficiency ratio of 30 per cent in 1939 to 53 per cent

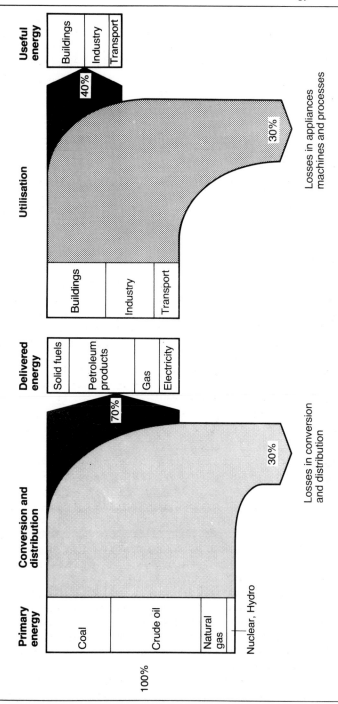

Fig. 6.1 United Kingdom Energy Flow (1975) (*Source:* Department of Energy, *Energy Paper No. 11,* HMSO 1976)

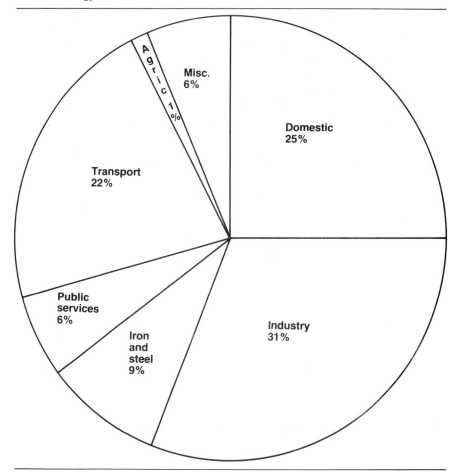

Fig. 6.2 Energy consumption by final users (1976) (*Source: Digest of Energy Statistics* 1977)

in 1975.[9] The increase in food prices since 1973, partly as a result of British entry into the EEC but also reflecting increased energy costs, has decreased demand, especially for pork products – a sector which had experienced considerable expansion in the two previous decades. Leach has shown that livestock farming, with the exception of sheep, is heavily dependent on energy inputs in the form of fertilisers, feed grains, machinery and fuels or electricity for heating purposes.[10] The pig and poultry sector, in particular, is a major energy user. In the good old days of cheap grain it made *economic* (if not ecological) sense to feed grain to animals rather than to use it directly as food. Two-thirds of the 15 million tons of cereals grown in the UK are fed to livestock and an additional 4 to 5 million tons of grain are imported to supplement supplies.[11] Animals are inefficient convertors of grain and emphasis on stock is mainly responsible for the energy intensiveness of modern farming. Pigs and poultry consume the greatest energy inputs, including 8 million tons a year of feed-grains.[12] With the era of cheap imported food now at an end, Slesser envisages further shortages in the future as developing nations increasingly keep their food at home rather than export it as a cash crop.

Alternative sources of feed will have to be found and, as costs increase, fodder and grass feed will regain their former importance in beef and dairy farming. Pigs and chickens do not digest grass but are capable of consuming human food waste. Dr Pereira, Chief Scientist at the Ministry of Agriculture, has estimated that 25 per cent of all food in Britain is wasted and much of this is table scraps which could be recycled as animal feed.[13] Hughes maintains that such a scheme is economically viable and that the separation and recovery of the useful food scraps is only slightly more expensive than the costs of disposal.[14]

Slesser envisages an expansion in home food production from the opening up of upland pastures which could yield grass-fed mutton, beef and venison in greater quantities than today.[15] Mellanby,[16] Allaby[17] and Blaxter[18] advocate a more radical approach, which would depend on a drastic change in the British diet. A shift in emphasis from livestock to crop production, a reduction in the use of inorganic fertilisers and a return to grass rather than barley feeding in animal husbandry could make Britain self-sufficient in food and also reduce fossil fuel energy inputs. The advocates of this policy would admit, however, that the present agricultural system could not quickly adapt to such a scheme. 'Its introduction would create financial havoc for farmers and the food industry unless the change to a system of this approximate type was slowly and carefully paced.'[19] The Government's White Paper in 1975 indicated no change of policy in this direction; indeed, much of the increase in production forecast in this document was to be achieved through the expansion of the pig and poultry sector and by increasing livestock densities.[20]

Industry

Industry consumes by far the greatest proportion of energy in the UK and the dramatic fall in total energy consumption in 1974 was mainly due to the three-day week and the consequent lack of demand for industrial fuel during the crisis. The industries consuming the greatest amounts of power are those which process basic materials; for example, the chemical industry is responsible for 14 per cent of energy used in industry and the iron and steel industry accounts for 9 per cent of total UK energy consumption (Fig. 6.2). Both industries have become progressively more energy efficient in the last decade – the chemical industry generates 30 per cent of its own power through combined heat and power systems[21] and the steel industry requires less energy inputs as a result of technological improvements in furnace design and the use of better quality ores. Basic material processing industries will face serious problems in the future, however, as raw material shortages lead to an escalation in costs. In energy terms, iron ore and other metallic minerals will need more energy per ton mined as exploration and development turns to leaner ores. Consequently, the recycling of redundant metal wastes will not only become economically viable but could also save vast quantities of energy. Table 6.1 shows the energy requirements for the production of metals and, while magnesium, aluminium and copper processing are the most energy consuming, considerable savings could be achieved there through recycling. Steel produced from scrap requires only 25 per cent of the energy compared to steel processes using ores.[22] In Britain around 20 million tonnes of ferrous scrap are recovered each year and

Dodds estimates that the use of scrap in the British Steel industry has saved the country from importing ore worth £900 million in the last 20 years.[23] The scrap utilisation process is more complex and unpredictable compared with the conventional pig iron method and Dodds suggests that countries well endowed with good quality ore would use this source of supply in preference to scrap. In the UK poor ore supplies have encouraged the greater use of scrap, with a beneficial effect on the import bill and the energy budget.

Table 6.1 Energy requirements for the production of metals (*Soure:* H. J. Alkema and E. V. Newland, Ch. 6, p. 99 in K. A. D. Inglis (ed.) *Energy: From Surplus to Scarcity,* Applied Science Publishers 1974)

	ton oil eq./ton
Magnesium	7
Aluminium	4–5
Recycle	1
Iron and steel	0.2–0.4
Copper	1–2
Recycle	0.2
Zinc	0.5

Not all waste materials can be recovered, either because of design complexities which result in high extraction costs or because of the prohibitive cost of collecting materials from dispersed sources. Simmons wryly remarks that lead can be reclaimed from exhausted vehicle batteries but not from lead bullets, except perhaps in Texas.[24]

One of the major problems which militates against energy saving measures is the growth in importance of the packaging and container industry. Consumer preference for 'convenience' goods is reflected in the time, money and energy spent by companies in the advertising and marketing of their products to increase sales. In many cases this leads to 'excessive' packaging to improve presentation and thus influence consumer choice. Packaging has a short life and presents a disposal problem, a problem which could be solved by a more efficient system of recycling waste paper and board. During the war, two-thirds of all waste paper and board was recycled; in 1977 only around a quarter was reclaimed despite our continuing dependence on imported wood pulp.[25] In 1976 Britain spent £3 million importing low-grade waste paper from the continent, especially from Germany, which has an efficient collection system. Imports in 1976 reached 100,000 tonnes, four times those of 1975 whilst local authority collections fell from 450,000 tonnes to 200,000 tonnes during the same period.[26] If the current rate of usage of paper is to be maintained, a greater proportion of waste will have to be recycled as world shortages of virgin fibre will occur in the 1990s.[27] It is impossible to recycle all paper produced – some is taken out of circulation more or less permanently in the form of books and wallpaper; also, paper cannot be recycled indefinitely as the fibres lose their strength on recycling and need to be reinforced by the addition of virgin fibres. Nevertheless, with the prospect of world shortages of wood pulp, Britain should place more emphasis on recycling and attempt to reverse the trends of recent years.

Unfortunately other materials used in the packaging industry do not justify recycling on economic grounds. The Oxfam Wastesaver scheme in Huddersfield discontinued its collection of glass, tin and plastics in 1977 because processing costs were barely covering the contracted prices. Walker feels that tin and plastics, although difficult to recycle economically, could become worthwhile with efficient collecting and processing.[28] Glass recycling, on the other hand, is hopelessly uneconomic. The Oxfam Wastesaver had sent Redfearns of Barnsley 275 million tonnes of glass by April 1977 – one-half of the total that the firm had received during the previous three years; however, processing costs alone were greater than the contracted price. Three years earlier, the City of York initiated a pilot scheme for the collection of waste glass with the cooperation of Redfearns, which has its headquarters in the city. The costs of collection and processing did not justify the continuance of the project. Unlike wood pulp for the industry, the raw materials for glass manufacture – sand, soda ash and limestone – are abundant in the UK and they only account for 13 per cent of the cost of the finished product.[29] Consequently, the cheapness and lack of scarcity of raw materials has not stimulated a move to recycling.

In the UK disposable glass containers outnumber returnables by 4:1,[30] and only 10 per cent of glass manufactured is re-used.[31] The success of the returnable milk bottle highlights the progress that could be made in moving away from our 'throw-away' society. In York, the pilot scheme for recycling glass suffered from a lack of public response after initial enthusiasm. The consumer must feel that the effort applied is worthwhile, hence, the delivery of milk has always been successful whereas returning 'odd' bottles may appear an aimless exercise for the average consumer. From a recycling viewpoint costs could be reduced significantly if all bottles were standardised into fewer types to facilitate separation and reprocessing. The marketing agencies, however, work against this principle in the hope that the colour and design of the bottle will influence consumer choice. But does the consumer, apart from the wine connoisseur or the chemist, really worry about the shape and colour of the bottle? If consumers accepted standardised liquid packaging, significant energy savings could be achieved.

Energy saving in industry

Few firms can boast that they 'saved it' before 1973. One exception is Boxfoldia, a Birmingham printed cartons manufacturer. As early as 1933 the firm was recording outside temperatures in order to ascertain whether heating was required inside the building. In 1947 cavities between the inner and outer roof linings were filled with 3½ inch fibreglass – a high standard of insulation even for the post-1973 era. Changes in boiler efficiency and other conservation measures have kept the companies' fuel costs to 1 per cent of the total selling cost of its products. Unfortunately, Boxfoldia is the exception rather than the rule; the Chief Scientist at the Department of Industry estimates that 5 per cent of the energy consumed by industry is wasted.

Benefits from conservation projects can be high: Marks and Spencer saved £4 million from 1974 to 1977 with their policy of 'good housekeeping'.[32] Courtaulds, experiencing difficult trading conditions, generated £10.5 million more cash in 1975 and 1976 with an energy conservation programme which has enabled them to save energy at a rate of 6 per cent a year.[33] The only major capital investment was the replacement of boilers; most measures were minor improvements such as draught-proofing, pipe lagging and frequent maintenance of all equipment to ensure that it was working at optimum efficiency. Dr Cunningham, Parliamentary Under-Secretary for Energy, opening the Insulation 77 exhibition at Wembley, stressed that a company was blunting its competitive edge if it wasted energy and illustrated his case with an example of Reckitt and Colman. They had spent £7,000 on lining a warehouse with 50 mm of insulation and were saving £5,000 a year in fuel costs; the £7,000 spent on pipe lagging would save £6,500 a year. It is clear from these figures that the payback period – under 18 months in both cases – could cut costs in industry. Whether industry responds depends on individual firms and their liquidity position. Insulation is, however, within the financial reach of most firms but the more elaborate schemes requiring large amounts of capital investment will need more careful consideration.

Domestic refuse recycling

Much of the energy consumed by the domestic sector (25 per cent of total energy consumption) is accounted for by space and water heating. Improvements in design, insulation and machine efficiency can help conservation of energy in the factory, the office and the home. Before outlining measures which can be adopted to achieve savings in the domestic sector, we can here continue the theme of recycling by considering the utilisation of domestic wastes either to produce heat or as reprocessable materials for industry. Such schemes would not only have ramifications in the field of energy conservation but, in addition, would lessen our reliance on imported raw materials and alleviate the problem faced by many local authorities in finding new waste disposal sites. On average, every member of the British public throws away 0.7 kilograms of household waste per day, which amounts to 15 million tonnes of rubbish per year for the total population.[34] At present 80 per cent of this waste is disposed of by controlled tipping and the remainder by incineration. The latter process has the advantage that it can reduce the amount of waste by as much as 90 per cent by volume and 60 per cent by weight, and authorities which have a lack of suitable tipping sites tend to prefer incineration in spite of the increased costs.[35]

The changing composition of domestic refuse in the last forty years is a reflection of an increasingly affluent society. Ash and cinders from coal fires were the dominant constituents of waste until the 1960s, when metals, glass and most importantly, paper begun to fill the nation's dustbins – a product of 'convenience' shopping and the packaging industry (Table 6.2).

Each year 4 million tonnes of paper, a million tonnes of ferrous metals, a similar quantity of glass, and tens of thousands of tonnes of non-ferrous metals are incinerated or tipped in the UK.[36] Domestic refuse is a potential source of re-usable material but the cost of separation and processing is the main drawback in the recovery of materials. For example, Oxfam Wastesaver in Kirklees have received cooperation from households in the separation of refuse into four separate containers – newspapers, mixed papers/magazines, glass/tins and rags/miscellaneous. As mentioned earlier, however, the costs of the containers and time and money involved in collecting and trasporting the refuse led to acute financial difficulties. Oxfam has been forced to rationalise: only paper and jumble is collected in September 1977, 47 workers were employed in

Table 6.2 The composition of domestic waste (%) (*Source:* J. Skitt, *Disposal of Refuse and other waste,* C. Knight & Co., 1972; and Waste Management Advisory Council, *1st Report,* HMSO 1976)

	1935	1968	1973
Fine dust and cinder	56.98	21.89	19.0
Vegetable and putrescible	13.71	17.61	18.0
Paper content	14.29	36.91	33.0
Metal	4.00	8.87	10.0
Rags	1.89	2.35	3.5
Glassware–jars, bottles cullet	3.36	9.11	10.0
Unclassified debris	5.77	2.14	5.0
Plastics	—	1.12	1.5

the scheme compared with 120 a year earlier (although the latter figure included 60 job creation employees).

The Warren Spring Laboratory in Hertfordshire, funded by the Department of Industry, has carried out considerable research into the technical and commercial feasibility of recovering materials from domestic waste in a form suitable for re-use in industry. The pilot refuse sorting plant at the laboratory produced sufficiently satisfactory results to encourage the Government to build Europe's first automatic waste recycling plant at Doncaster, due to be completed in 1979. The laboratory utilises refuse from Stevenage and has shown that 50 per cent by weight – a greater proportion by volume – can be recovered for use as low grade fuel or as substitute raw materials in manufacturing or construction industries. Table 6.3 summarises the possible products that can be recycled from domestic wastes.

Refuse is generally high in calorific value and can be used as a source of energy production. Nevertheless, most local authorities construct incinerators to reduce the amount of waste and recover neither heat nor material for reprocessing.[37] In Nottingham the local authority has an efficient heat recovery scheme. When the project was first conceived in the 1960s, it was planned to coincide with the redevelopment of the city centre in order that heat from the incinerator and an old industrial power station could be channelled into new offices, shops and dwellings. The scheme provides heat for two shopping centres, covered markets, major public buildings, educational establishments and thousands of corporation homes.

Heat recovery from incineration does present difficulties. The flow of refuse is not constant, resulting in problems of tailoring supply to demand. If breakdowns occur within the plant, standby power is necessary. The most important consideration, however, is that of finance, and the siting of the incinerator is crucial if heat recovery is to be possible. Nottingham is a model example as the incinerator was part of a comprehensive city-centre development. In most cases, however, the erection of a new incinerator is not part of an integrated planned development. Consequently, the costs involved in converting the existing infrastructure to a waste heat scheme would be prohibitively high in many cities.

Another possibility for the use of domestic waste as a form of heat is pyrolysis. This involves the heating of waste in the absence of oxygen to yield fuel and char. Different processes are being investigated in the USA but the

Table 6.3 Recoverable products from domestic waste (*Source:* Warren Spring Laboratory: *The New Prospectors* 1976)

Product	Typical/ Approximate weight as % of feed	Description	Comments
Combustibles	30	Mainly paper, plastic film and textiles	Possible use as supplementary fuel in existing solid fuel boilers?
Paper-rich fraction	10	Mainly (»85%) paper but plastic film, textiles, feathers and fine dust also present	Additional processing required before acceptable for use in board making.
Tin-plate product	5	Mainly (»95%) discarded tin cans with adhering labels, lacquer linings and traces of original contents	Commercially acceptable to iron founders. Could be shredded and cleaned for detinning.
'Massive iron'	1	Discarded kitchen utensils, household goods, car components etc.	Commercially acceptable as low-grade scrap iron.
Non-ferrous metal concentrates	«1	Mainly aluminium but zinc and copper also present	Recovery and cleaning circuits under development.
Glass-rich fraction	6	Overwhelming preponderence of white glass but amber and green glass also present	Suitable for use in the manufacture of green bottle glass.
Putrescible-rich fraction	20	Vegetable and animal food wastes, wood, plastic, paper and glass	Possibly usable as feed-stock for methane generation or for production of single cell protein for incorporation in animal feed-stocks.

Warren Spring Laboratory is operating a pilot scheme which it is hoped can ultimately produce the heat equivalent of 100,000 tons of coal from every million tons of domestic refuse.[38]

Energy conservation in the home

Energy savings in the home can be made in two ways, either by improving the heating efficiency of appliances and processes in supplying heat or by minimising the use of heat at source by wasting as little as possible. From 1960 to 1972 the quantity of energy supplied to domestic households remained constant although the number of households increased by 18 per cent.[39] This was a result of improved efficiency: better insulation standards in newer buildings, which often have lower ceilings than existing housing stock, in addition to improvements in the efficiency of appliances.

As over 85 per cent of the heat supplied to households is used for space and water heating, the choice of heating equipment can minimise heat loss. Natural gas-fired central heating is more efficient than other forms of space heating,[40] but the homeowner's decision to convert to gas is fraught with uncertainties. How long will he remain in the same house? If he moves will he recoup this expenditure? What is the payback period to guarantee a return on the capital outlay? It may be better to keep the existing system but improve insulation in order to use less fuel. Some families are too poor to contemplate a change of heating system. Around 7 million tonnes of coal a year continue to be burned in open fires which are less than 20 per cent efficient.[41]

Research at Aston University has demonstrated that open fires can radiate 50 per cent of the heat released if a fluidised bed system is used and efficiencies of over 80 per cent have been achieved by adding surface heat exchangers to the flue. This is an important development because of coal's long term potential compared with other fossil fuels. Improvements in the efficiency of appliances will save energy but, unfortunately, the most efficient space heating system (natural gas) will have only a limited lifespan. By 2000 overall mean efficiencies will be lower than at present because of increasing conversion losses in the secondary fuel industries, especially the electricity supply industry, which will take up the slack in domestic energy demand as natural gas reserves dwindle.

The Building Research Establishment has estimated that an ultimate saving of over 15 per cent of primary energy consumption could be achieved in buildings without impairing comfort.[42] Simple thermal insulation measures – cavity fill, double glazing and loft insulation – could cut consumption by 3 to 4 per cent. Clearly a wholesale programme of thermal insulation would lead initially to an increase in energy consumption through the extra production of materials such as glass, glassfibre and cavity foam, but the average British home is so badly insulated that the insulation measures would pay for themselves in fuel terms within a few years. It is estimated that energy worth £500 million a year is being wasted because of inadequate insulation.[43] One half of the homes with accessible lofts have no insulation at all; two-thirds of the council houses and one third of privately owned houses are uninsulated; in the private rental sector, only one quarter of the properties are insulated. Boyle has estimated that to bring the 19 million homes in Britain up to high standards of thermal insulation would create employment for 4,000 workers per year for 30 years or, if a crash programme was undertaken 12,000 per year for 10 years.[44]

The harsh reality is that all houses in Britain will not be insulated despite the cost-effectiveness of all forms of insulation. Loft insulation pays for itself in two years, draught stripping in three, wall insulation in five and a half years and double glazing, the least cost-effective, has a payback period of ten years.[45] The capital cost involved, however, is high if all types of insulation are undertaken – the above measures, for example, carried out in a typical post-war semi would cost £1,300. It is possible that a settled home owner may invest in his home by extending his mortgage but council tenants and landlords are unlikely to spend this amount of money on their houses.

The potential of CHP schemes and the heat pump

In the future Combined Heat and Power schemes (CHP) and the use of heat pumps could make a significant impact in the domestic and commercial heating markets in addition to contributing to the conservation of energy. 300 private CHP schemes are already in operation in the UK and 20 per cent of industry's requirements for electricity is met in this manner.[46] Combined heat and power, however, 'provides a good example of the difference between maximum theoretical efficiency and optimum economic efficiency'.[47] CHP schemes can achieve a theoretical thermal efficiency of 85 per cent under ideal conditions.[48] This high efficiency is achieved by utilising a by-product from power generation – wasted hot water. This water, however, is of too low a temperature to be used directly for industrial or district heating schems and requires upgrading, with a consequent loss of electricity production. Further problems are encountered in the distribution of this heat, as future electricity generation is envisaged to be based on nuclear plants which are sited at some distance from areas of demand. The flexible demand for heat and power is one of the main factors which affects the theoretical 85 per cent efficiency of such schemes. The electricity industry's constant problem is the maintenance of expensive plant which lies idle most of the time and meets only peak demand. The supply of heat linked to electricity generation compounds the problem, especially in the summer months when electricity is still required but heat requirements are much less. The CEGB policy of the 'merit order' system whereby new efficient power stations supply the base load and older plants 'top up' supply during peak periods conflicts with the principle of total energy systems based on CHP plants which must operate continuously. In practice a new station when first commissioned tends to operate day and night until it is overtaken by the introduction of more efficient plant; therefore, as it ages a CHP station produces electricity increasingly inefficiently in order to continue supplying heat to its district heating customers. Because of the imbalance between supply and demand, the National Economic Development Office estimates that the thermal efficiency of CHP schemes is more likely to be 45 per cent in practice, although the report points out that this figure compares favourably with the overall mean efficiency for household energy use – 38 per cent in 1975.[49] A working party assessing the costs of conversion of power stations and developing district heat network concluded that CHP schemes could not compete with conventional heating methods, where a supply infrastructure already exists.[50] An exercise by the SSEB using an obsolete power station at Pinkston near the centre of Glasgow, gives a practical illustration of the economics of CHP operations. Conventional forms of electricity generation were considered more cost competitive, although the 'heat' component of the scheme would give an economic return on investment over a long payback period.[51]

Even if we accept the argument that the success of district heating schemes on the continent is not sufficient justification in itself for the UK to develop along similar lines, an intermediate solution may be possible which strikes a balance between conserving the heat wasted in power generation and the economic problems of its utilisation. The Department of Energy accepts that decisions have to be taken well in advance for CHP schemes because of the long lead times required to install the plant and its associated networks. It is estimated that one-quarter of the UK homes could be converted to district heating in 25 years, which would coincide with the fall-off of natural gas production from the North Sea. A possible solution would be to convert city power stations, such as the Glasgow Pinkston station, to CHP plants. City centre plants tend to be obsolescent, less efficient and therefore low rated in the merit order, but their main advantage is that they are close to high density loads such as public buildings, offices and flats located in central areas. Furthermore, the gradual build-up of such a system would provide long-term employment in the boiler making and generator supplier industries. With many European countries seriously interested in district heating, the export potential of CHP schemes of smaller size – 50 to 250 MW – may be considerable. The modification of city stations would be coupled with maintaining efficiency in larger plants. In the long term this approach could lead to the electricity supply industry re-capturing its lost markets, especially the space heating market, from gas.

For district heating to be totally successful, fundamental changes in life-style would be necessary. High density loads are essential. The advantage which France and other continental countries have over the UK for the implementation of CHP schemes is their acceptance of apartment living whereas in the UK lower density housing is preferred. This means that in Britain, under present conditions, the development of the heat pump, rather than district heating schemes is indicated. The Building Research Establishment and the Electricity Council advocate the further development of heat pumps for water and space heating. The Council's Capenhurst Research Centre, after experiments with methods of saving energy in conventional British houses, chose a high standard of thermal insulation and the use of heat pumps as the two best methods to achieve savings.[52] The Building Research Establishment claim that energy consumption would have been 7 to 9 per cent less in the UK if present domestic space heating and hot water requirements had been met by the use of heat pumps.[53]

A heat pump extracts heat from a low-temperature heat source – air or water – and upgrades this heat to a higher temperature for use within a building. It operates on a similar principle to the domestic refrigerator; the cooling, however, takes place outside the building. The coefficient of performance (COP) of heat pumps is between 2 and 3, that is units will deliver 2 to 3 kW of useful heat for each kW of electricity consumed. The advantages of this technology are that greater flexibility in the siting of power stations would be possible compared with district heating schemes, and efficiencies are comparable with other conventional forms of heating. The balance will probably favour heat pumps in the future as improvements in design could lead to a COP of 4 to 5 and they also can be used in waste heat recovery in industry and the home.

Why has the acceptance of the heat pump been limited in the UK? Much depends on existing heating methods – as long as the UK has sufficient natural gas supplies to meet space heating demand, cost considerations weigh against heat pumps making a major impact in this market. At present the actual as opposed to the theoretical COP is around 2, and capital costs are much greater than for conventional heaters.[54] Heat pumps are increasingly used in the USA because they can also be used as air conditioners, and this has helped to reduce the effective capital cost. It is likely that heat pumps will play an important role in space and water heating towards the end of the century, when oil and gas supplies fail to keep pace with demand. This 'breathing space' will enable development to improve heat pump performance which could lead to mass production, and a reduction in the overall capital cost. In the UK electric heat pumps were due to be marketed in 1978 by Lennox Industries Limited at a wholesale cost of £110 and an installed cost of £200, a figure which is competitive with conventional systems.[55]

Most of this discussion has stressed the limitations of existing housing stock in the prevention of heat loss. The revised Building Regulations, introduced in January 1975, roughly doubled insulation standards in the construction of new buildings and, although a step in the right direction, greater consideration could have been given to other aspects of building design and construction. The *Architect's Journal* questions the emphasis on thermal insulation, not as a means of saving energy but to increase comfort in the home.[56] The Department of Energy, in a circular to Scottish local authorities, stresses that 'design teams should be encouraged to consider the "energy use" of all new buildings from the earliest planning stage'.[57] The NEDO report states that occupiers of houses with a north-facing living-room are liable to pay 10 per cent more for their heating than if their living-room was facing south.[58] No doubt, each individual new house could be meticulously designed to cater for its specific needs but, as a general rule, greater emphasis could surely be given to the shape of the house, its degree of exposure to the elements and the types of construction materials used. Conflict between aesthetic and energy values will arise; for example, the problem of the north-facing aspect with beautiful views which merits a large picture window but which consequently accelerates heat loss is not easily resolved. Unfortunately, most modern buildings – industrial premises, office blocks and housing estates – have a uniformity of design regardless of location, whether it be Inverness or Brighton. According to Burberry, the overall design for energy conservation is prevented by the structure of the building design professions because of the division in responsibilities between the architects who design the shell of the building and the heating engineers who design the heating equipment.[59] This lack of coordination has exercised the RIBA Council, which advocates that architects be educated to design for conservation and laments the tendency for research finance to be directed to the transport, rather than the building sector.[60] This observation is emphasised by the poor level of representation of the architectural profession on the Advisory Council on Energy Conservation whose reports have been biased towards transport issues.

Transport

In 1976 the transport sector consumed 22 per cent of total British energy,[61] energy which is almost exclusively derived from petroleum products. Of the 71.5 million tonnes of petroleum delivered for energy use in 1976, 40 per cent was consumed by the transport sector. The stability of oil prices from 1974 to 1978 was reflected at the petrol pump, where the oil companies conducted price wars to reduce mounting stocks. This situation has proved short-lived. Even at present consumption levels, oil resources will become exceedingly scarce by the end of the century and oil can be more efficiently used in other sectors, for example space heating, than in the internal combustion engine, where only 15 to 20 per cent of the energy supplied goes towards moving the vehicle. It would seem logical to advocate a shift from road to alternative less prodigal methods of transport for the movement of people and goods but it is undeniable that this inefficient mode of transport remains the most popular. Both freight and private passenger road transport have increased their share of the total volume of trips generated at the expense of other sectors. (Tables 6.4 and 6.5). The watershed year for passenger transport in the UK was 1955, the last year when public transport was more important than the private car. By 1975, four-fifths of the

Table 6.4 Passenger transport in Great Britain ('000 m. passenger kilometres) (*Source:* C. G. Bamford and H. Robinson *Geography of Transport,* MacDonald and Evans, 1978.

	Road		Rail	Air
	Buses and Coaches	Private Transport		(including N. Ireland and Channel Islands)
1954	80	76	39	0.3
1955	80	87	38	0.3
1960	71	144	40	0.8
1965	63	233	35	1.7
1970	56	306	36	2.0
1975	54	357	35	2.2
Average annual % change				
1955-65	–2.4	+10.4	–0.8	+16.5
1965-75	–1.5	+ 4.4	—	+ 2.5

Table 6.5 Goods transport in Great Britain ('000 m. tonne–kilometres) (*Source:* C. G. Bamford and H. Robinson *Geography of Transport,* MacDonald and Evans, 1978.

	Road	Rail	Coastal shipping	Inland waterways	Pipeline
1955	37.0	34.4	20.4	0.3	0.2
1960	48.4	30.1	15.3	0.3	0.3
1965	68.8	25.2	25.0	0.2	1.3
1970	85.0	26.8	23.2	0.1	2.9
1975	91.8	23.5	18.3	0.1	3.3
Average annual % change					
1965-75	+2.9	–0.7	–3.1	–10.2	+9.6

passenger/kilometres travelled were by private car and two-thirds of the freight transported was moved by road. To reverse the trends of the last 20 years will require a radical change in Government policy. The next section will show that the Government's White Paper on Transport Policy is moving towards a more radical approach. It seems unlikely, however, that the above pattern will alter significantly in the next 15 years. Once again, energy efficiency does not necessarily relate to cost efficiency. Masefield has shown that in terms of passenger miles per imperial gallon of petroleum products the hierarchy of energy efficient modes is:

1. Inter-city bus
2. British Rail inter-city train
3. London Transport bus
4. Underground train
5. Private car
6. Boeing 747
7. British Airways (Concorde)

The costs per mile were very dependent on the load factor, that is the number of passengers travelling by each mode of transport.[62] For example, if the seats on London Transport buses and trains were continuously half filled, they would make profits at current fares.[63] Public transport may be more energy efficient but it is only optimally utilised at peak periods in the morning and evening. The Government could price motorists off the road through a policy of motor taxation whilst subsidising public transport to an even greater degree to ensure that cost and energy efficiency are in phase. Such measures would, however, be very unpopular with the public.

When it is considered that 60 per cent of all journeys to work are less than 5 miles,[64] there is considerable scope for the provision of cycle lanes to assist safe cycling. The price mechanism has led to a move towards two-wheel rather than four-wheel mobility and 1975 broke a 16-year record for motorcycle sales.[65] More expensive fuel has caused civil aviation and shipping companies to reassess their position. The number of scheduled flights has been reduced to secure higher load factors – the Laker skytrain is a good example of a reduction of costs by filling each plane to capacity. In shipping, cruising speeds were reduced, but in some cases this led to a drop off in carrying capacity and normal speeds were reintroduced.

The Advisory Council on Energy Conservation has devoted most of its energy papers to identifying methods of reducing energy consumption.[66] Whilst acknowledging that a move from road to rail is desirable in private and freight transport, the papers are primarily concerned with modifications to the existing modal pattern. With present technology, fuel can be saved by improving the efficiency of the petrol engine or by moving towards the more efficient diesel engine in all forms of road transport. The reduction of vehicle weight, improvements in carburettor and ignition systems and the compulsory use of radial tyres would all contribute to reducing fuel consumption. Many drivers are now willing to sacrifice power for economy and the high-performance private cars has become less dominant since 1973. Chapman has shown that from 1960 to 1968 the average size of registered cars in the UK fell from 1400 cc to 1300 cc, partly because of the success of the Mini; from 1968 to 1973 the trend was reversed.[67] The move by the car industry to introduce new models in the economy range since 1973 has again reversed the trend in consumer preference. In Europe Ford, which had fostered a high-performance image through its involvement in rallying and motor racing, has invested heavily in the Fiesta, an 'economy' model. Similarly Vauxhall, a company traditionally in the larger car market, introduced the small engine capacity Chevette in 1975. This trend will accentuate as petrol prices begin to move upwards and the car industry continues to emphasise economy in its advertising campaigns.

The motor car industry has undertaken most of the minor modifications recommended by the Advisory Council on Energy Conservation in addition to investing in the development of alternative fuels and new drive systems, such as the external combustion engine (stirling engine) or electrical power systems. The large amounts of capital invested in existing engine assembly lines make it impossible for any car manufacturer to discard existing production units. As a result, every major European car manufacturer is working on small diesel engine projects in order to achieve a built-in flexibility in engine lines by switching from petrol to diesel engines and vice versa. Diesel engines for motor cars have relatively poor performance characteristics, and their size, weight and noisiness have limited their market potential with the exception of larger vehicles – over 2-litre engine capacity – which have been selling well for Mercedes and Peugeot.

Ford UK have devised a sonic-idler carburettor which vaporises fuel more effectively, producing better combustion with reduced fuel consumption. Other manufacturers, including Saab, have renewed their interest in turbocharged engines, which allow more oxygen into the petrol mixture to give better fuel combustion. Another Advisory Council recommendation which is being put into practice is the production of lighter vehicles. Aluminium engines were rare 10 years ago but are now being increasingly adopted by manufacturers such as Peugeot, Volvo and Renault. Masefield prophesies that car bodies will also be made of aluminium in the future because of its lightness and rust-free properties.[68] Although perhaps aesthetically unappealing, the use of lighter plastic compounds in external and internal fitments reduces the amount of weight propelled by the engine. Plastic bumpers and less sophisticated and lighter internal fascias, are common features in modern cars.

From an energy efficient viewpoint oil would be better utilised in the future for premium uses, for example, in the chemical industry than for transport. With the increasing cost competitiveness of electricity from nuclear or alternative energy generation, a shift back to public transport seems likely with further electrification of railways and perhaps a comeback for the trolleybus and the tram. For aircraft the fuel of the future could be liquid hydrogen. Although the Department of Energy feels that the difficulties of storage and possible leakage problems would preclude the development of liquid hydrogen as a fuel in car transport, it has been suggested that a supersonic aircraft fuelled on liquid hydrogen would be 40 per cent cheaper to build than its present day equivalent.[69] The flexibility of the private car will ensure its continued popularity despite its relative high cost and inefficiency. Research and development into alternative transport fuels – methanol, synthetic hydrocarbons and electric traction – will probably guarantee personal private mobility into the twenty-first century. The success of electric traction – the most likely alternative – depends on the development of suitable batteries. The lead acid battery is inadequate for electric traction and can only be used in short delivery vehicles with a range of 50 miles and a top speed of 50 mph. To develop a battery power supply which can give a greater speed and range requires a breakthrough in the type of battery with a high energy density, for example, the sodium sulphur battery, the likely successor to the low density lead acid battery.

The Government response

The approach of the Government towards the conservation of energy has been to advise and encourage individual consumers to 'save it' rather than to adopt a rigorous policy which would influence consumer choice through subsidisation or taxation policies. In energy supply, the Secretary of State for Energy has argued that the Government has a central role to play but in energy use 'the Government acting on its own cannot hope to provide a substitute for consumers taking the necessary and right decisions.'[70] The Government has set an example for other companies to follow. The Department of Environment's Property Services Agency (PSA) is attempting to achieve a 30 per cent cut in fuel consumption from 1974 to 1979. This target is on schedule with the PSA surveying and converting buildings to computerised heating control systems which match the supply and demand of heat to the intermittently-used PSA buildings. The simple measure of reducing temperatures to 65°F – 3°F below the ceiling recommended by the Department of Energy – is saving millions of pounds per year. The nationalised fuel industries have also emphasised conservation within their organisation and in their advice to consumers. British Gas have gone one step further by establishing a School of Fuel Management at Solihull in the Midlands to advise different sectors of management on how to utilise gas more efficiently.

The Government is exerting its influence in transport policy and the proposals incorporated in the 1977 White Paper[71] showed significant changes from those advocated in the 1976 Consultation Document.[72] Expenditure on public transport was not reduced as planned; instead, less revenue was to be allocated to new road construction. Clearly, public expenditure cuts were mainly responsible for this revision of policy, but it can be surmised that the Advisory Council's three energy papers on transport may have influenced the change of direction. The business car owner has also come under the scrutiny of the Council, which recommended a tightening up of the incentives given to those users. The Council points out that 90 per cent of business cars are used for private use to some extent and that non-business mileage accounts for 4 to 8 per cent of the total mileage of business cars each year.[73] As a result, the Inland Revenue began to tax car expenses over a certain level for 'essential' users, which should help to trim some of this excess mileage.

The price mechanism has played a significant role in the conservation of energy, with the private motorist bearing the brunt of not only raw material price increases but also the increase caused by 25 per cent VAT on motor spirit imposed in November 1974. This budget also began the phasing out of subsidies to the nationalised fuel industries to create a true pattern of energy use related to costs of production. From October 1973 to January 1977 domestic coal prices rose by 95 per cent compared with the increases of 65 per cent for gas, 110 per cent for oil and 135 per cent for electricity during the same period. In the industrial sector, price increases were more marked, coal prices rising by 110 per cent, gas by 180 per cent, oil by 280 per cent and electricity by 130 per cent. Increases in the price of fuels obviously influence demand; however, the role of other factors, such as the economic recession or the weather, makes an assessment of price elasticities a formidable task. Nevertheless it has been estimated that savings of up to 6 per cent, worth around £600 million, may have been made in 1975.[74]

In 1974 the prospect of two-tier petrol rationing, a three-day working week and electricity cuts promoted a sense of urgency to save energy. The 12-point plan of December 1974 complemented the budget proposals of the previous month. Heating and lighting restrictions were imposed on non-domestic buildings and thermal insulation standards were imposed in the construction of new houses. In transportation, a decision to reduce further the lead content of petrol was deferred, speed limits were lowered on motorways and dual carriageways and oil price increases were to be weighted to fall proportionally more on motor spirit. In industry, loans were to be made available for energy saving programmes, and the Government indicated that legislation would be enacted if firms did not include in their annual accounts the cost savings which could be attributed to energy conservation measures. This package, which also included the inauguration of the 'Save It' campaign, was intended to save £100 million on Britain's oil deficit of £3,446 million in 1974. The proposals were modest but constituted a starting point from which to build a more rigorous energy conservation policy. Much of the urgency of the crisis years has disappeared. The publicity surrounding North Sea oil and gas has undermined the value of the commendable 'Save It' campaign. In the three years to March 1978 the Government spent £8 million to advertise conservation on television, in the press and through leaflets and posters advising how to save energy in the home and at work. Much of the publicity has been directed towards the domestic sector, perhaps in the hope that top-level managers may implement at work policies that they find worth while at home. The main problems in concentrating attention on publicity are, firstly, that consumers inevitably reach a saturation point and the intial impact will wear off and, secondly, that an information and advisory service tends to reach those sections of the community which are seeking to increase their level of comfort rather than save energy.

In industry the Energy Thrift and Energy Audit schemes are intended to supply commercial firms with a greater range of information on energy saving measures. By gaining experience from one-day visits to firms which volunteered to be part of the schemes, the Department of Energy hopes to identify areas in need of further research in addition to using this information in future 'Save It' campaigns. In terms of financial benefits to industry, the Government has been far from generous. The loan scheme is unsuccessful partly because interest rates are on a par with commercial finance. Smaller companies tend to have only limited sources of finance for such investment. One barrier to commercial investment in energy saving measures has stemmed from the inclusion of fuel savings in the annual accounts because until August 1976 the Price Code ruled out the carrying forward of improved profits attributed to energy savings. The only cases in which grants can be received for insulation and other related energy improvements to buildings are firms in the Assisted Areas which undertake structural alterations.

Government conservation policy was highly criticised by the Select Committee on Science and Technology in July 1975.[75] The Committee was concerned with the lack of progress towards a more positive approach to energy conservation, estimating that a cut of 15 per cent in energy consumption could be made without sacrificing output, employment or standards of living, at a saving of £1,000 million.[76] The Committee was critical of the present administrative machinery, including the part-time Advisory Council on Energy Conservation, and recommended that a 'task force' of full-time officials reporting directly to the Prime Minister should be responsible for national policy on energy saving. Altogether the Committee made 42 recommendations. It suggested the replacement of the industry loan scheme by a larger loan scheme under which grants and loans to industry would be conditional on the installation of 'best practice fuel using plant and equipment'.[77] It proposed test studies severely restricting car use in selected cities, and checks on fuel efficiency as part of the Department of the Environment tests. It also recommended that gas and electricity appliances should carry information on their rating and that the Government should inform the public of the most cost-effective way to use them. In the fuel supply industries further consideration was urged for inverted tariffs (under which the consumer pays proportionally more, not less, as at present, for additional consumption). In electricity generation the Committee advised that the use of oil in power stations should be reconsidered.

The Government was slow to respond to the Select Committee report, but eventually rejected most of the recommendations.[78] Instead of creating a 'task force', the Government set up a committee of ministers to promote and co-ordinate energy conservation. The Government again emphasised that it was not going to take consumers' decisions for them and that the orientation of policy would continue to be to the provision of more publicity, information and advice. To help the smaller firm the Government would subsidise up to half the cost (to a maximum of £30) of a one-day visit by an energy consultant.

In September 1977 the Energy Secretary announced the first part of a revitalised energy conservation programme which included the doubling of the subsidy for the survey scheme to £60, further demonstration projects to emphasise waste heat recovery, a free advisory service to industry, commerce and the public sector, and more studies on the problems and energy saving potential of 21 energy-intensive industries. Three months later a £320-million package was launched with an aim of saving £700 million within a decade. Most of the investment will be concentrated in the public sector to improve thermal insulation and heating controls in public buildings.[79]

The programme, which also includes further consultation with the motor industry to improve fuel consumption, follows a similar pattern to its predecessors. The Government hopes that if it sets an example, the private home owner and industry alike will follow its lead and use the advice services created to promote energy savings. It can be argued, as it was by the Select Committee, that Government legislation should enforce energy conservation. Interference with consumer choice, however, is a politically difficult issue and the action taken by the Government is a reflection of this.

While there is no easy option for the Government, an important omission is nonetheless evident in the approach of both the Government and the Select Committee in that no major measures are envisaged to promote change in the present energy-intensive economy. The new direction in transport policy is encouraging, but other sectors, including agriculture, exhibit trends which appear to involve greater rather than less energy use. As economic recovery in Britain continues, much of the impetus has gone from energy saving projects such as waste recycling schemes. The example of the Oxfam Wastesaver is indicative of the influence of economic forces over energy efficiency motives in the short term. Similarly, since the creation of the Waste Management Advisory Council, Britain has imported more wastepaper and local authority collections continue to fall. With impending raw material shortages by the end of the century, it is imperative that the Government supports recycling schemes – they save not only energy but also valuable foreign exchange.

Notes and references

1. Department of Energy (1976a) *Energy Research and Development in the United Kingdom, Energy Paper No. 11*, HMSO, p. 28, and *Energy Policy Review, Energy Paper No. 22*, HMSO (1977a).

2. A. Porteous (1975) in J. Lenihan and W. W. Fletcher (eds) *Energy Resources and the Environment*, Ch. 3, Blackie, pp. 78, 79.

3. Department of Energy (1977b) *District Heating Combined with Electricity Generation in the United Kingdom, Energy Paper No. 20*, HMSO, p. 29.

4. P. Chapman (1975) *Fuel's Paradise: Energy options for Britain*, Penguin, pp. 163–5.

5. T. Beresford (1975) *We Plough the Fields, British Farming Today*, especially Ch. 6, Penguin.

6. G. Leach (1976) *Energy and Food Production*, IPC Science and Technology Press, p. 21.

7. M. Slesser (1975) in J. Lenihan and W. W. Fletcher (eds) *Food Agriculture and the Environment*, Ch. 1, Blackie, pp. 1, 2.

8. Leach (1976) op. cit., p. 31.

9. K. Blaxter (1975) 'Can Britain feed herself?', *New Scientist*, 20 March 1975; and Beresford (1975) op. cit., p. 82.

10. Leach (1976) op. cit., pp. 98, 99.

11. K. Laidlaw (1975) *The Party's Over*, World Development Movement, p. 14.

12. Ibid., p. 14.

13. Laidlaw (1975) op. cit., p. 20.

14. D. Hughes 'Waste not, want not', *New Scientist*, 20 March 1975.

15. Slesser (1975) in Lenihan and Fletcher, op. cit., p. 19.

16. K. Mellanby (1975) *Can Britain Feed Itself?*, Merlin Press.

17. M. Allaby (1975) 'Can we feed ourselves?', *Ecologist,* July 1975.

18. Blaxter (1975) op. cit.

19. Leach (1976) op. cit., p. 36.

20. *Food From Our Own Resources*, Cmnd 6020, HMSO, 1975.

21. *Financial Times*, 24 January 1977.

22. Porteous (1975) in Leniham and Fletcher op. cit., p. 76.

23. M. J. Dodds (1973) 'Scrap Planning', TRP8, Department of Town and Regional Planning, University of Sheffield, p. 23.

24. J. G. Simmons (1974) *The Ecology of Natural Resources*; Edward Arnold, p. 271.

25. *Guardian*, 'Special Report on Recycling', 13 October 1977.

26. *Financial Times*, 17 March 1977.

27. The *Guardian* Special Report highlights the usage of paper by taking the production of its own newspaper as an example. Every issue of the *Guardian* consumes 40 tonnes of paper, each tonne represents 17 fully-grown trees, 125 kg of sulphur, 160 kg of limestone, 250,000 litres of water, 4,000 kg of steam and 255 kilowatt hours of electricity.

28. J. Walker (1977) 'Trouble at Mill', *Undercurrents*, No. 23, August 1977.

29. *War on Waste: A Policy for Reclamation,* Cmnd 5727, HMSO, 1974, p. 20.

30. Porteous (1975) in Lenihan and Fletcher, op. cit., p. 85.

31. *Guardian* Special Report, op. cit.

32. Advertisement in the *Financial Times* Supplement 'Managing Energy', 26 September 1977.

33. Ibid.

34. Warren Spring Laboratory (1976) *The New Prospectors*, HMSO.

35. Incineration costs between £10 to £12 a ton compared to £1.25p to £2 a ton for controlled tipping.

36. E. Douglas and P. R. Birch (1976) 'Recovery of potentially re-usable materials from domestic refuse by physical sorting', *Resource Recovery and Conservation* 1, 1976.

37. In Derby, 6 per cent of the ballast is metal and this is reclaimed magnetically.

38. Warren Spring Laboratory leaflet, 'Domestic and Industrial Wastes'.

39. National Economic Development Office (1974) *Energy Conservation in the United Kingdom; Achievements, Aims and Options,* HMSO, p. 26.

40. See ibid; p. 28. and Economic Commission for Europe (1976) *Increased Energy Economy and Efficiency in the ECE Region*, United Nations, p. 71.

41. *Financial Times*, 28 January 1975.

42. Building Research Establishment (1975) *Energy Conservation: a study of energy consumption in buildings and possible means of saving energy in housing*, BRE, p. 1.

43. Mr Charles Ryder, Head of Energy Conservation Technology at the Department of Energy, quoted in the *Financial Times*, 15 October 1977.

44. G. Boyle (1976) 'Let's have some more radioactivity', *Undercurrents* No. 15, April 1976.

45. G. Leach (1977) 'Saving it', *Undercurrents* No. 25, December 1977.

46. Department of Energy (1976a) op. cit., p. 74.

47. Eric Varley (1974) *Towards the Efficient Use of Energy*, HMSO, p. 7.

48. National Economic Development Office (1974) op. cit., p. 34.

49. Ibid.

50. Department of Energy (1977b) op. cit., p. 29.

51. South of Scotland Electricity Board (1976) *Central Glasgow District Heating Study*, April 1975–March 1976, SSEB.

52. *Energy World*, No. 38, June 1977, pp. 8, 9.

53. Building Research Establishment (1975) op. cit., p. 22.

54. Unfortunately, the COP falls as the temperature outside the building drops and this is when demand is greatest.

55. Leach (1977) op. cit.

56. The editorial in the *Architect's Journal*, 11 September 1974, p. 592.

57. Department of Energy (1976b) *Energy Conservation (Scotland)*, Circular 1/76, 25 February 1976, para. 12.

58. National Economic Development Office (1974) op. cit., p. 35.

59. P. Burberry 'Conserving energy in buildings', *Architect's Journal*, 11 September 1974, p. 616.

60. Editorial in *Architect's Journal*, 12 October 1977, p. 690.

61. Road transport accounted for 17 per cent of total energy consumption in 1976.

62. Sir Peter G. Masefield (1976) 'The challenge of change in transport', *Geography* **61**, November 1976, p. 213.

63. Ibid., p. 219.

64. National Economic Development Office (1974) op. cit., p. 44.

65. *Financial Times*, 22 January 1976.

66. Advisory Council on Energy Conservation *Paper 2: Passenger Transport: short and medium term considerations, Energy Paper No. 10*, HMSO, 1976; *Paper 5: Road Vehicle and Engine Design: short and medium term energy considerations, Energy Paper No. 18*, HMSO 1977; *Paper 6: Freight Transport: short and medium term considerations, Energy Paper No. 24*, HMSO, 1977.

67. Chapman (1975) op. cit., pp. 223, 224.

68. Masefield (1976) op. cit., p. 217.

69. Department of Energy (1976a) op. cit., p. 84.

70. Varley (1974) op. cit., p. 3.

71. *Transport Policy*, Cmnd 6836, HMSO, 1977.

72. *Transport Policy: A Consultation Document* (2 volumes), HMSO, 1976.

73. Advisory Council on Energy Conservation *Paper 2*, op. cit., *Energy Paper No. 10*, p. 12.

74. Department of Energy (1977a) *Report of the Working Group in Energy Elasticities, Energy Paper No. 17*, HMSO, p. 54.

75. Select Committee on Science and Technology, First Report (1975) *Energy Conservation*, HMSO.

76. Ibid., pp. 14, 15.

77. Ibid., p. 29.

78. *Energy Conservation* Cmnd 6575, HMSO, 1976.

79. Local authorities will be allocated £114 million through housing subsidies to improve the standard of thermal insulation in 2 million council homes.

Summary

Britain is relatively more fortunate than other industrialised countries in that she is well endowed with energy resources. With abundant coal, a well-established nuclear industry, geographically advantageous sites for the further development of wave, wind and tidal power, and the short-term benefits of offshore oil and gas, Britain could be self-sufficient in energy into the twenty-first century.

The Government has to achieve the right mix of fuels for the best use of these energy resources. It is to be hoped that the open debate on energy policy initiated by Mr Benn while Energy Secretary will quickly give rise to a much more flexible policy than has been followed in the past. Successive governments in the 1950s and 1960s adopted a 'coordination through competition' policy which has proved unsuccessful because of their failure to anticipate future world energy trends. Hence, the NCB was encouraged to invest in an expansion programme in the 1950s only to be subsidised in the 1960s after being obliged to prune its excess capacity because of competition from cheap oil supplies. During the rapid transformation to an oil-based economy, dangers of political action to curtail supplies were underestimated in spite of warnings throughout the 1960s and early 1970s of the likelihood of oil supply restrictions by OPEC nations.

The 'energy crisis' of 1973/74 was in essence a political crisis; however, a real economic crisis due to scarcity of energy supplies will occur in the late 1990s unless appropriate action is taken now. The main obstacle to constructive action to avert this situation is that unless a crisis appears imminent, the public will not be convinced of the necessity to pay high prices for their energy today in order to preserve it for future generations. The Government has therefore kept open as many options as possible to retain more flexibility of supplies than in the 1960s. The Arab oil embargo of 1973/74 could be a blessing in disguise since it highlighted the wastefulness of energy in our society and led to the introduction of conservation measures by the Government. Alternative energy resources, especially renewable energy, began to figure more prominently in Research and Development expenditure whereas these energy forms would have been overlooked if the cheap oil era had continued. As a result, throughout the 1970s, the public was gradually made more aware than previously of energy decisions which affect their everyday lives.

Britain is a small, densely populated island and there is inevitable conflict between the fuel supply industries and the communities which will be affected by a decision to construct a new coal mine, LPG or oil terminal or a nuclear power station. In most instances, the public are willing to accept a greater consumption of energy but not at the expense of their immediate environment.

The decision to appoint an independent Standing Commission to advise on energy policy and its interaction with the environment is an important break-through because of the contribution it can make to informing public discussion on the national interest in the balance between energy and the environment.

It is unrealistic to define a precise energy strategy because of the complex nature of the relationship between energy and other industries. For example, the decision to bring forward the order for the Drax B power station, although preserving jobs in the power manufacturing industries, will have repercussions in the coal mining and electricity industries. A long-term policy must be framed in such a way as to take account of social, economic and environmental consider-ations. In the short term the main problem is not one of energy shortage but of energy surplus. The present sluggish demand for electricity and the ordering of Drax B and two further AGRs has produced an oversupply situation which will continue into the 1980s. This means that the coal and nuclear industries – envisaged by the Government as playing a dominant role in long-term energy planning – could have excess capacity for domestic demand up until 1985. Offshore oil and gas will be operating at peak production during the same period and will therefore aggravate the oversupply situation.

Even with a controlled production depletion policy, however, oversupply could eventually give way to a shortage and thus necessitate the import of scarcer, high-priced supplies in the 1990s.

In the light of the evidence produced for each of the fuel industries what pathway should we take for future energy planning? Energy conservation should enjoy top priority in any strategy. The Government has made a useful start in the implementation of an energy conservation programme by its advertising campaigns to encourage energy management in home, office and factory and through financial assistance to the private and public sector to improve thermal insulation. Greater attention could be focused on the conservation potential of waste materials and waste heat. At present the recycling of household and industrial wastes is not an economic proposition as the Oxfam Wastesaver project found to its cost. However, the long-term potential benefits of recycling could be considerable, both in terms of saving energy and conserving raw materials. Waste heat from incinerators and power stations could also play a significant role in saving fuel in a long-term strategy, although the cost competitiveness of natural gas would seem to preclude the economic implementation of such schemes until the late 1980s.

The development of offshore oil and gas resources has given the UK sufficient breathing space to assess all of the energy options available in a long-term strategy. If energy conservation measures begin to take effect by the 1980s and the price of oil continues to rise faster than inflation, it is possible that a third phase of oil development will occur in the western offshore zones, involving the exploitation of marginal discoveries in the North Sea. If further associated gas deposits are also discovered and developed, and the plans to tap reserves through a gas gathering pipeline are carried through, overall reserves could meet domestic demand until the first quarter of the twenty-first century; oil production levels could be sustained throughout the 1980s and perhaps into the 1990s, depending on the depletion policies imposed by the Government.

The 'energy gap' by 2000 may be negligible if the scarcity of world oil supplies encourages further development of offshore resources and a more efficient utilisation of energy onshore reduces consumption. Nevertheless, coal, nuclear and renewable energy technologies must be developed to guarantee long-term supplies. However, the pace of development of each of these technologies poses problems. Expansion of mining capacity will be necessary to meet demand by 2000, but in the short term this will result in coal being stock-piled unless electricity demand increases at a faster rate than it has done throughout the 1970s. The further development of fluidised bed combustion and SNG technologies would give the NCB greater market flexibility and decrease its dependence on the power station market. The nuclear industry, on the other hand, is tied exclusively to the electricity market and by the next century the development of this form of power will release coal supplies for non-electricity uses.

The experience of the nuclear industry does not inspire confidence in the benefits of direct investment in the construction of additional power stations in the short-term. With the exception of the Magnox design, other reactors have not proved successful. The delays in the AGR programme are attributed to technical problems but their likely replacement, the US designed PWRs, are unreliable and concern for the safety of the emergency core cooling system, the subject of a public hearing in 1972, ultimately led to the disbanding of the Atomic Energy Commission and a slowdown in nuclear reactor construction. The decline in demand for electricity and hence nuclear power in the short term may give the industry adequate time to develop a safe design of proven reliability before commitments are made to the more controversial fast breeder reactor. By the late 1980s we shall have a clearer picture of the future availability of uranium supplies on which a decision on FBR commercial development may be based.

A delay in the development of nuclear power will enable more research to be carried out into renewable energy resources. Improvements in solar panel technology should make it cost-competitive in the 1980s whilst large scale wind and wave power prototypes may supply power to the grid by the mid to late 1980s. By 2000 enough experience will have been accumulated about these technologies to leave open a choice about which combination of power plant– nuclear, renewable and coal-fired – should provide base load electricity in the twenty-first century.

Environmental considerations should be taken more into account in a future energy strategy. The production, transportation and utilisation of energy has detrimental effects on the environment and not enough attention has been given to minimising this disruption. The dereliction created by coal mines constructed in the pre-1947 period must not be allowed to recur in the last quarter of this century when new mines, often in rural sites, will have to be opened to meet increased demand by 2000. Selby is a model example of an environmental mine and it is to be hoped that the NCB will continue to plan to limit environmental impact.

Oil and gas have created environmental problems both offshore and onshore. Only the Orkney and Shetland authorities have satisfactorily attempted to control oil-related developments by taking statutory powers to ensure that the local population receives the full benefits of the oil boom. Offshore, more adequate precautions are required to tackle blowouts, platform damage and pipeline failure, if and when they occur. The increased risks of oil pollution from the transhipment of North Sea oil could be substantially reduced through stricter policing of 'rogue' captains who break the laws of the sea and the imposition of high fines on shipping companies which operate sub-standard ships run by inadequately trained crews.

Nuclear power has always been associated with environmental risk because of its military connotations. Reprocessing technology presents the major dangers. Judging by evidence produced at the Windscale inquiry, radioactive emissions from low-active wastes need more thorough monitoring. Until vitrification technology is perfected, the storage of high-active waste poses a moral problem because of its potential effects on future generations. The export of nuclear technology could lead to a proliferation of nuclear weapons as present international safeguards do not seem to prevent countries with nuclear aspirations from converting 'peaceful' plants to weapons technology. The most quoted alternative to nuclear power – renewable energy – is not free from environmental problems. The tidal power barrage scheme across the Severn Estuary requires the incorporation of pollution and silt control measures before it becomes environmentally acceptable. The scale of the planned wind and wave power plants would create visual intrusions into the landscape and offshore generators may cause navigational problems in addition to modifying coastal ecology because of the removal of energy from the sea.

A strategy for the future must therefore be based on coal and involve the conservation of oil and gas to provide a stopgap until nuclear and renewable technologies are sufficiently developed to make a greater contribution to electricity supplies. All types of energy have associated environmental problems and in such a small island as Britain, it will become increasingly important to ensure that the future development of energy resources does not cause irretrievable damage to the environment.

Index